SpringerBriefs in Computer Science

For further volumes:
http://www.springer.com/series/10028

Amit Vasudevan · Jonathan M. McCune
James Newsome

Trustworthy Execution on Mobile Devices

Amit Vasudevan
James Newsome
Carnegie Mellon University
Pittsburgh, PA
USA

Jonathan M. McCune
CyLab, Carnegie Mellon University
Pittsburgh, PA
USA

ISSN 2191-5768   ISSN 2191-5776   (electronic)
ISBN 978-1-4614-8189-8   ISBN 978-1-4614-8190-4   (eBook)
DOI 10.1007/978-1-4614-8190-4
Springer New York Heidelberg Dordrecht London

Library of Congress Control Number: 2013941912
ARM, the ARM Logo and any other trademark found on the ARM Trademarks List that are referred to or displayed in this book are trademark[s] or registered trademark[s] of ARM Ltd or its subsidiaries. Other names may be trademarks of their respective owners.

Printed on acid-free paper

Springer is part of Springer Science+Business Media (www.springer.com)

To my loving wife Deepa and my dear sons Arjun and Akshay

—Amit Vasudevan

Kathleen, Adeline, and Lillian

—Jonathan M. McCune

For my loving wife, Bonnie Bogovich

—James Newsome

Preface

In June 2012, we presented a paper entitled *Trustworthy Execution on Mobile Devices: What Security Properties Can My Mobile Platform Give Me?* at the 5th International Conference on Trust and Trustworthy Computing [59]. Subsequently, we were invited to expand our paper with the belief that given the increasing importance of mobile device security, our review of the current state of the art on trusted execution on mobile devices would be a great help to the security community, particularly to graduate students. This inspired us to expand our original paper into the form you see here. We hope that it will be of service to the security and privacy community.

Arlington, VA, March 2013   Amit Vasudevan
Santa Clara, CA   Jonathan M. McCune
Pittsburgh, PA   James Newsome

Acknowledgments

The authors are especially grateful to our collaborators, Emmanuel Owusu, Zongwei Zhou, Virgil Gligor, and Adrian Perrig, whose insights and enthusiasm greatly enriched our work.

The authors would also like to thank Bill Hohl and Joe Bungo at ARM for generously providing technical information and research hardware and software to support this work.

This research was supported by CyLab at Carnegie Mellon University (CMU), Northrup Grumman Corp., and Google Inc. The views and conclusions contained here are those of the authors and should not be interpreted as necessarily representing the official policies or endorsements, either express or implied, of CyLab, CMU, Northrup Grumman Corp., Google Inc., or the U.S. Government or any of its agencies.

Contents

1 Introduction 1

2 Demand for Applications Requiring Hardware Security 5

3 Desired Security Features. 9
   3.1 Isolated Execution. 9
   3.2 Secure Storage 10
   3.3 Remote Attestation 11
   3.4 Secure Provisioning. 12
   3.5 Trusted Path 13

4 Available Hardware Primitives. 15
   4.1 ARM Platform: Hardware and Security Architecture 15
   4.2 Isolated Execution. 16
      4.2.1 Split-World-Based Isolated Execution 16
      4.2.2 Virtualization-Based Isolated Execution. 19
   4.3 Secure Storage 21
      4.3.1 Secure Elements 21
   4.4 Remote Attestation 22
   4.5 Secure Provisioning. 22
   4.6 Trusted Path 23
   4.7 Design Gaps and Challenges. 23
   4.8 Platform Case Studies 24
      4.8.1 ARM Versatile Express 24
      4.8.2 FreeScale i.MX53 27
      4.8.3 Texas Instruments OMAP^TM and M-Shield^TM 30
      4.8.4 Samsung Exynos^TM. 34

5 Isolated Execution Environments 37
   5.1 Parallel Isolated Execution 37
   5.2 Hypervisors 38
      5.2.1 Hypervisor Attributes for Mobile Devices 38

    5.3   Hypervisor Case Studies.   39
          5.3.1   KVM/ARM   40
          5.3.2   CodeZero.   41
          5.3.3   OKL4   42
          5.3.4   EmbeddedXen   44
          5.3.5   Xen/ARM   45
          5.3.6   eXtensible Modular Hypervisor Framework   46
    5.4   Discussion   47
          5.4.1   Limitations of Paravirtualization.   48

6   API Architectures   49
    6.1   API Types   49
    6.2   App-IEE-Only Model Versus App-IEE + Module-IEE Model . .   49
    6.3   Candidate APIs.   50
          6.3.1   Mobile Trusted Module.   51
          6.3.2   OnBoard Credentials   51
          6.3.3   TrustZone^{TM} API   52
          6.3.4   GP Trusted Execution Environment   53

7   Analysis and Recommendations   55
    7.1   Research Community Recommendations   56
    7.2   Application Developer Recommendations.   56
    7.3   Platform Integrator Recommendations   57
    7.4   Hardware Vendor Recommendations   57

8   Summary.   59

References   61

About the Author   65

Curriculum Vitae   67

Acronyms

AES   Advanced Encryption Standard
API   Application Programming Interface
COTS   Commodity Off-The-Shelf
DES   Data Encryption Standard
DMA   Direct Memory Access
MSSF   Mobile Simplified Security Framework
NFC   Near Field Communication
NVRAM   Non-Volatile Random Access Memory
OEM   Original Equipment Manufacturer
OS   Operating System
PKI   Public-Key Infrastructure
POS   Point-Of-Sale
RSA   Rivest Shamir Adleman Algorithm
RTS   Root-of-Trust for Storage
SD   Secure Digital
SE   Secure Elements
SHA   Secure Hash Algorithm
TCB   Trusted Computing Base
TrEE   Trusted Execution Environment
TCM   Tightly Coupled Memory
TPM   Trusted Platform Module
UICC   Universal Integrated Circuit Card
USIM   Universal Subscriber Identity Module

Chapter 1
Introduction

We are putting ever more trust in mobile devices. We use them for e-commerce and banking, whether through a web browser or specialized *apps*. Such apps hold high-value credentials and process sensitive data that need to be protected.

Meanwhile, mobile phone Operating Systems (OS) are untrustworthy. While in principle they attempt to be more secure than desktop OSes (e.g., by preventing modified OSes from booting, by using safer languages, or by sandboxing mechanisms for third-party apps such as capabilities), in practice they are still fraught with vulnerabilities.

Mobile OSes are as complex as desktop OSes. Isolation and sandboxing provided by the OS is routinely broken, c.f. Apple™ iOS jail-breaking by *clicking a button on a web page* [16, 50]. Mobile OSes often share code with open-source OSes such as GNU Linux™, but often lag behind in applying security fixes, meaning that attackers need only look at recent patches to the open-source code to find vulnerabilities in the mobile device's code. Therefore, there is a need for isolation and security primitives exposed to application developers in such a way that they need not trust the host OS.

We argue that this problem is severe enough to have garnered significant attention outside of the security community. Demand for mobile applications with stronger security requirements has given rise to add-on hardware with stronger security properties (Chap. 2). This situation is unfortunate, given that many current mobile devices already have hardware support for isolated execution environments and other security features. However, these features are not made available to all parties who may benefit from their presence.

Today's mobile device hardware and software ecosystem consists of multiple *stake-holders*, primarily comprising the OEM (handset manufacturer), telecommunications provider or carrier, application developers, and the device's owner (the human user). Carriers typically also serve in the role of platform integrator, customizing an OEM's handset with additional features and branding (typically via firmware or custom apps). To date, security properties desirable from the perspectives of application developers and users have been secondary concerns to the OEMs and carriers [14, 37, 56].

A. Vasudevan et al., *Trustworthy Execution on Mobile Devices*, SpringerBriefs in Computer Science, DOI: 10.1007/978-1-4614-8190-4_1, © The Author(s) 2014

The historically closed partnerships between OEMs and carriers have lead to a monolithic trust model within today's fielded hardware security primitives. Everything "inside" is assumed to be trustworthy, i.e., the software modules executing in the isolated environment often reside in each other's trusted computing base (TCB). As long as this situation persists, OEMs and carriers will not allow third-party code to leverage these features. Only in a few cases, where the OEM has partnered with a third party, are these features used to protect the *user's* data (c.f. Chap. 2, Google Wallet).

We approach this scenario optimistically, and argue that there is room to meet the needs of application developers and users while adding negligible cost. We thus define the principal challenge for the technical community: **to present sound technical evidence that application developers and users can simultaneously benefit from hardware security features without detracting from the security properties required for the OEMs and carriers.**[1] Our goal in this paper is to systematize deployed (or readily available) hardware security features, and to provide an extensive and realistic evaluation of existing (largely academic) proposals for multiplexing these features amongst *all* stake-holders.

We first discuss the demand for mobile applications with stronger security requirements that has given rise to add-on hardware with stronger security properties (Chap. 2).

We then proceed in Chap. 3 by defining a set of security features that may be useful for application developers that need to process sensitive data. Our focus is on protecting secrets belonging to the *user*, such as credentials used to authenticate to online services and locally cached data.

We next provide an overview of hardware security features available on today's mobile platforms (Chap. 4). We show that hardware security features that can provide the desired properties to application developers are prevalent, but they are typically not accessible in COTS devices' default configurations.

We then move on to evaluate existing proposals (given the hardware security features available on mobile devices) for creating a trustworthy execution environment that is able to safely run sensitive applications that are potentially considered untrustworthy by other stake-holders (Chap. 5).

We show that multiplexing these secure execution environments for mutually-distrusting sensitive applications is quite possible if the threat model for application developers and users is primarily software-based attacks (Chap. 6).

Finally (Chap. 7), we provide an end-to-end analysis and recommendations for the current best practices for making the most of mobile hardware-based security features, from the points of view of each stake-holder.

Unfortunately, without firmware or software changes by OEMs and carriers, individual application developers today have little opportunity to leverage the

---

[1] We wish to distinguish this challenge from proposals that OEMs increase their hardware costs by including additional hardware security features that are exclusively of interest to application developers and users. Our intention in this book is to emphasize practicality, and thus define such proposals to be out of scope.

hardware security primitives in today's mobile platforms. The only real options are either to partner with a mobile platform integrator, to distribute a customized peripheral (e.g., a smart-card-like device that can integrate with a phone, such as a storage card with additional functionality), or to purchase unlocked development hardware. We provide recommendations for OEMs and carriers for how they can make hardware-based security capabilities more readily accessible to application developers without compromising the security of their existing uses.

Chapter 2
Demand for Applications Requiring Hardware Security

Application developers are an important driver of sales for the entire smartphone industry since compelling mobile applications and services are at the forefront of a user's smartphone experience. Handset manufacturers and mobile OS designers have undertaken many initiatives to encourage application developers to create apps for their platforms. For example, OEMs have provided development environments, technical support, and publishing and sales infrastructures.

Does providing third-party developers with access to hardware-supported security features make sense for the OEMs or carriers? This is an important consideration for an industry where a few cents in cost savings can be the deciding factor for features. We show that there are many applications on mobile devices that require strong security features, and that must currently work-around the lack of those features. Being forced to deal with these work-arounds stifles the market for security-sensitive mobile applications, and endangers the security of the applications that are deployed anyways.

We detail several applications requiring specific hardware security features.

Google Wallet™[1] allows consumers to use their mobile phones as a virtual wallet. The application stores users' payment credentials locally, which are then used to make transactions via near field communication (NFC) with point-of-sale (POS) devices. To store the users' credentials securely, Wallet relies on a co-processor called a Secure Element (SE) which provides isolated execution (Sect. 3.1), secure storage (Sect. 3.2), and a trusted path (Sect. 3.5) to the on-board NFC radio. Unfortunately, the SE only runs code that is signed by the device manufacturer.[2] This may be because the SE lacks the ability to isolate authorized modules from each-other, or it may simply be considered a waste of time. As a result, developers without Google™'s clout will not be able to leverage these capabilities for their own applications.

Google Wallet™ achieves security, but it is not an open platform. It relies on a co-processor called a Secure Element (SE), where only trusted programs are allowed

---

[1] http://www.google.com/wallet/how-it-works-security.html

[2] http://www.google.com/wallet/how-it-works-security.html

A. Vasudevan et al., *Trustworthy Execution on Mobile Devices*, SpringerBriefs in Computer Science, DOI: 10.1007/978-1-4614-8190-4_2, © The Author(s) 2014

to run. Additionally, the details on the exact security properties of the Secure Element are thin.[3]

There is evidence that Apple™ has similar plans for its products; they recently published a patent for an embedded SE with space allocated for both a Universal Subscriber Identity Module (USIM) application and "other" applications [49].

Traditionally, setting up a merchant system required making arrangements with a bank or credit card processing company in a process that involved fees and could take weeks or months to complete. Mobile POS products promise to bypass the complicated setup process and allow vendors to make sales anywhere.

Services such as Square and GoPay allows merchants to complete credit card transactions with their mobile device using an application and a magnetic stripe reader [38]. While Square's security policies[4] indicate that they do not store credit card data on the mobile device, the data does not appear to be adequately protected when it passes through the mobile device. Researchers have verified that the stripe reader does not protect the secrecy or integrity of the read-data [41]. This implies that malware on the mobile device could likely eavesdrop on credit-card data for swiped cards or inject stolen credit-card information to make a purchase [41].

These applications could benefit greatly from the hardware-backed security features we describe in Chap. 3. A trusted path (Sect. 3.5) could enforce that the intended client application has exclusive access to the audio port (with which the card readers interface), thus protecting the secrecy and integrity of that data from malware. They could also benefit greatly from a remote attestation mechanism (Sect. 3.3) which the servers could use to ensure that received-data is actually from the authorized client-application, and that it used a trusted-path to the reader, thus helping to ensure that the physical credit card was actually present.

Companies have attempted to fill the gap left behind by the lack of developer-accessible hardware security features on mobile devices, by implementing removable Secure Elements (SEs). Removable Secure Elements (also known as Independent Secure Elements or Secure Memory Cards) are SEs interfaced to removable memory such as a Universal Integrated Circuit Card (UICC) or Secure Digital (SD) Card [45].

Secure memory cards allow third-party developers to develop applications for one Secure Element interface instead of having to account for the specific interface requirements of various handset manufacturers. Since secure memory cards are removable, consumers can easily move credentials to other handsets or devices. On the other hand, removable SEs may be more vulnerable to physical attack (e.g., the SE may be more easily lost, stolen, or corrupted).

UICC is a general-purpose platform for smart-card applications. UICC is capable of hosting applications for the card issuer—such as USIM for voice and data access—in addition to hosting non-telecommunications applications including mobile payments and ticketing. As a multi-tenant SE, the UICC's operating system manages memory access for multiple mutually-distrusting applications [29].

---

[3] http://www.google.com/wallet/how-it-works-security.html

[4] https://squareup.com/security

Giesecke & Devrient offer the Mobile Security Card which is a 2 GB microSD card coupled with a Secure Element—the Mobile Security Card. The SE on the Mobile Security Card supports cryptographic functions including SHA-256, DES/3-DES, AES, and RSA.[5] This functionality enables security-sensitive applications such as disk encryption, single sign-on, building access control, and PKI key management. Tyfone offers a similar product called SideSafe.[6]

The number of new applications requiring hardware security features is evidence that there is demand for hardware-backed security primitives among third-party businesses and application developers. Unfortunately, some of the workarounds may actually increase the attack surface for fraud [41]. Several third parties have stepped in to provide hardware-backed security features in the form of removable Secure Elements. OEMs could provide a more tightly integrated experience for developers, and avoid potential security vulnerabilities by opening up pre-existing hardware security primitives to application developers.

---

[5] http://www.gd-sfs.com/the-mobile-security-card/mobile-security-card-se-1-0/

[6] http://tyfone.com

Chapter 3
Desired Security Features

Here we describe a set of features intended to enable secure execution on mobile devices. This can be interpreted as the wish-list for a security-conscious application developer. The strength of these features can typically be measured by the size, complexity, and attack surface of the components that must be relied upon for a given security property to hold. This is often referred to as the *trusted computing base* (TCB).

On many systems, the OS provides security-relevant APIs for application developers. However, this places the OS in the TCB, meaning that a compromised OS voids the relevant security properties. We briefly discuss whether and how the security features below are provided on today's mobile platforms, and some strategies for providing these properties to applications without including the OS in the TCB.

3.1 Isolated Execution

Isolated execution gives the application developer the ability to run a software module in complete isolation from other code. It provides secrecy and integrity of that module's code and data at *run-time*. Today's mobile OSes provide process-based isolation to protect applications' address spaces and other system resources. However, these mechanisms are circumventable when the OS itself is compromised.

To provide isolated execution that does not depend on the operating system, some alternative execution environment not under control of the OS is required. Such an environment could be provided by a layer running under the OS on the same hardware (i.e., a hypervisor), or in a parallel environment (such as a separate coprocessor). We examine some candidate isolated execution environments and their suitability for mobile platforms in Chap. 5.

Regarding today's mobile platforms, the Meego Linux™ distribution for mobile devices does include provisions for isolated execution. Meego's Mobile Simplified

A. Vasudevan et al., *Trustworthy Execution on Mobile Devices*, SpringerBriefs
in Computer Science, DOI: 10.1007/978-1-4614-8190-4_3, © The Author(s) 2014

Security Framework (MSSF) implements a trusted execution environment (TrEE) that is protected from the OS [33]. However, this environment is not open to third party developers.

3.2 Secure Storage

Secure storage provides secrecy, integrity, and/or freshness for a software module's data *at rest* (primarily when the device is powered off, but also under certain conditions based upon which software has loaded). The most common example demonstrating the need for secure storage is access credentials, such as a cached password or a private asymmetric key. Other examples include sensitive information cached for offline consumption, such as bills or medical information.

Most mobile OSes provide this property at least using file system permissions, which are enforced by the operating system. File system permissions alone can be circumvented by compromising the OS itself. They can also be circumvented by offline attacks such as by loading an alternative OS that does not respect those permissions, or by removing the storage media and accessing it directly.

A stronger form of secure storage can be built using a storage location that is physically protected, and with access control implemented independently of the OS. E.g., on the PC, the TPM has a small amount of on-board NVRAM for this purpose. A physically protected piece of secure storage used in this way is called a *root of trust for storage*, or RTS.

A root of trust for storage can be used to bootstrap a larger secure storage mechanism, using *sealed storage*. The sealed storage primitive uses a key protected by the RTS to encrypt the given data, and to protect the authenticity of that data and of attached meta-data. The metadata includes an access-control-policy for which code is authorized to request decryption (e.g., represented as a hash over the code), and potentially other data such as which software module sealed the data in the first place. Sealed data (ciphertext) can then be stored on an unprotected storage device.

Extra steps are required to provide *state continuity* for sealed data on untrusted storage; otherwise the sealed data may be undetectably rolled back to an older version. This could be a problem, e.g., if the sealed data is an access-control list or a revocation list. Freshness protection can be implemented using trusted counters, a small piece of protected storage for a counter or hash, or a trusted time source [43].

Symbian and Meego make use of protected memory and sealed storage [33]. Symbian's installer system distinguishes between removable and permanently-installed storage, and it calculates a hash over any applications installed to removable media, and stores that hash in permanently-installed storage. Applications executed from removable media are subsequently integrity-checked using the relevant hash.

MSSF uses keys kept in its Trusted Execution Environment (TrEE) (Sect. 3.1) to protect the integrity of application binaries, and to provide a sealed storage facility, which *is* available to third party developers [33]. While this offers protection against offline attacks, since third party applications are not allowed to execute in the TrEE, data protected by this mechanism is vulnerable to online attacks via a compromised OS.

Recent versions of iOS combine a user-secret with a protected device-key to implement secure storage [6]. However, the device-key does not appear to be access-controlled by code identity, meaning that an attacker can defeat this mechanism if he is able to obtain the user secret, e.g., via malware, or via performing an online brute-force attack [24, 31].

Android™ offers an AccountManager API [2]. The model used by this API supports code modules that perform operations on the stored credential rather than releasing them directly, which would make it amenable to a model with sealed storage and isolated execution. Unfortunately, it appears that the data is currently stored in plaintext, and can be retrieved via direct access to the storage device or by compromising the operating system [1, 64].

Android™ began offering file-system encryption in version 3.0 [3]. However, this feature is protected only by a user-secret entered at boot time, meaning that it can be circumvented by compromising the operating system at runtime, or by brute-forcing the user-secret in an offline attack. Android™ 4.0 will also support a new keychain API [4]. The details of this API and of how the data is protected are not yet available.

3.3 Remote Attestation

Remote attestation allows remote parties to verify that a particular message originated from a particular software module. Remote attestation is useful in cases where a remote service wishes to ensure that it is communicating with a known client, and not with malware. For attestation to be meaningful, it must attest to the entire TCB of the given application. For an application running on a normal OS, the attestation would necessarily include a measurement of the OS kernel, which is part of that TCB, and of the application itself. A remote party, such as an online banking service, could use this information, if it knew a list of valid OS kernel identities and a list of valid client banking-app identities, to ensure that the system had booted a known-good kernel, and that the OS had launched a known-good version of the client banking app.

Remote attestations are more meaningful when the TCB is relatively small and stable. In the example of a banking application, if a critical component of the app ran as a module in an isolated execution environment with a remote-attestation capability, then the attestation would only need to include a measurement of the smaller isolated execution environment code, and of the given module. Not only would it be easier to keep track of a list of known-good images (assuming that the isolated execution environment's code is relatively stable), but the attestation would be more meaningful because the isolated execution environment is presumed to be less susceptible to run-time compromise. This is important because the attestation only tells the verifier what code was *loaded*; it would not detect if a run-time exploit overwrote that code with unauthorized code.

Attestation mechanisms are typically built using a private key that is only accessible by a small TCB (Sect. 3.1) and kept in secure storage (Sect. 3.2). A certificate issued by a trusted party, such as the device manufacturer, certifies that the corresponding public key belongs to the device. One or more platform configuration registers store measurements of loaded code. The private key can then be used to generate signed attestations about its state or the state of the rest of the system.

Security-conscious developers of such application ecosystems can benefit from the ability for a mobile device to *report* on its health. An obvious example is a client-side anti-malware application that wishes to report that the most recent scan completed successfully and using the latest known signatures. However, a whitelist-based scenario is where such an architecture can really shine.

Given support for isolated execution (Sect. 3.1) and secure storage (Sect. 3.2), a remote attestation protocol can be used to allow a remote server to deterministically ascertain that the intended client-side code has loaded. This is especially pertinent to the many mobile device applications that use a "cloud-based" architecture, where the long-term storage of users' data takes place in a data center, and attestation can be used as an additional authentication metric to ensure that the desired client application has loaded. This whitelist-based capability frees the application developer from the burden of trying to understand all of the other third-party code that might be present and potentially malicious on the mobile device (because that code is no longer a part of the TCB for the sensitive portion of the application).

It can be useful to contrast an attestation scheme to a *secure boot* scheme. Secure boot is the process of performing integrity checks (e.g., verifying a cryptographic hash or digital signature) on each stage of the boot process, and halting if any stage fails its check. A fully booted device is thus implicitly believed to be in an approved configuration. Attestation separates the process of measuring (performing a cryptographic hash) each stage of execution from the process of evaluating whether a set of measurements represents a valid configuration. Especially when there are multiple stake-holders, secure boot does not scale all the way to individual third-party applications. Remote attestation can convey meaningful information under such conditions, because individual attestations can be sent to the relevant stake-holders for evaluation.

Some forms of remote attestation are implemented and used on today's mobile platforms [33]. However, as far as we know, no such mechanisms are made available to arbitrary third-party developers.

3.4 Secure Provisioning

Secure provisioning is a mechanism to send data to a *specific software module*, running on a *specific device*, while protecting that data's secrecy and integrity. This is useful for migrating data between a user's devices. For example, a user may have a credential database that he wishes to migrate or synchronize across devices while

ensuring that only the corresponding credential-application running on the intended
destination device will be able to access that data.

One way to build a secure provisioning mechanism is to use remote attestation
(Sect. 3.3) to attest that a public encryption key belongs to a particular software
module running on a particular device. The sender can then use that key to protect
data to be sent to the target software module on the target device.

Some of today's mobile platforms implement mechanisms to authenticate external information from the hardware stake-holders (e.g., software updates), with the
hash of the public portion of the signing key stored immutably on the device [33].
Other secure provisioning mechanisms are likely implemented and used by device
manufacturers to implement features such as digital rights management. As far as
we know, however, secure provisioning mechanisms are not available for direct use
by arbitrary third-party developers on mobile platforms.

3.5 Trusted Path

Trusted path protects authenticity, and optionally secrecy and availability, of communication between a software module and a peripheral (e.g., keyboard or touchscreen)
[26, 30, 36, 57, 66]. When used with human-interface devices, this property allows
a human user to ascertain precisely the application with which she is currently inter-
acting. With full trusted path support, malicious applications that attempt to spoof
legitimate applications by creating identical-looking user interfaces will conceivably
become ineffective. While human factors abound in designing the precise UI elements [18, 48], the technical underpinnings that enable any such architecture remain
a significant challenge.

Trusted path to sensors and actuators can be another useful feature. For example,
trusted paths to sensors can be used to facilitate "citizen-journalism" applications,
where a software module uses trusted path to ensure that it is receiving unaltered
sensor input, and then uses remote attestation mechanisms to attest to the accuracy
of the sensed data [25, 47].

Building secure trusted paths is a challenging problem. Zhou et al. propose a
trusted path on commodity x86 computers with a minimal TCB [66]. Their system
enables users to verify the states and configurations of one or more trusted-paths
using a simple, secret-less, hand-held device. In principle, many mobile platforms
support a form of trusted path, but the TCB is relatively large and untrustworthy.
For example, the *Home* button on iOS and Android™ devices constitutes a *secure
attention sequence* that by design uncircumventably transfers control of the user
interface to the OS's "Home" screen. Once there, the user can transfer control to the
desired application.

However, the TCB for such mechanisms includes the entire OS. For the Android™
Home button, the TCB also includes third-party apps that the user installed with the
"launcher" capability. An app that the user installs with the launcher capability can
replace the Home screen with its own arbitrary interface (potentially impersonating

the previous Home screen if it didn't advertise itself as a custom Home screen). It is then free to impersonate other apps by spoofing the UI of the requested app instead of launching the requested app.

The OS can be removed from the TCB of such trusted paths by preventing the OS from communicating directly with the device and running the device driver in an isolated environment. This requires the platform to support a low-level access-control policy for access to peripherals. ARM's TrustZone™ extensions facilitate this type of isolation (Sect. 4.2.1).

Chapter 4
Available Hardware Primitives

In this chapter we discuss currently-available hardware security primitives with a focus on existing smartphone and tablet platforms. As the vast majority of these platforms are built for the ARM architecture, we first present a generic ARM platform hardware and security architecture, focusing our discussion on platform hardware components that help realize the features discussed in Chap. 3. We then identify design gaps and implementation challenges in off-the-shelf mobile devices that prevent third-party application developers from fully realizing the desired security properties. Finally, we provide two case studies of inexpensive mobile *development* platforms with myriad security features, to serve as references against which to compare mass-market devices.

4.1 ARM Platform: Hardware and Security Architecture

ARM's platform architecture comprises the Advanced Microcontroller Bus Architecture (AMBA) and different types of interconnects, controllers and peripherals. ARM calls these the "CoreLink", which has four major components (Fig. 4.1).

- *Network interconnects* are the low-level physical on-chip interconnection primitives that bind various system components together. These include switches, bridges, and routing fabric. AMBA defines two basic types of interconnects: (i) the Advanced eXtensible Interface (AXI)—a high performance master and slave interconnect interface, and (ii) the Advanced Peripheral Bus (APB)—a low-bandwidth interface to peripherals.
- *Memory controllers* correspond to the predominant memory types: (i) static memory controllers (SMC) interfaced with SRAM, and (ii) dynamic memory controllers (DMC) interfaced with DRAM.
- *System controllers* include the: (i) Generic interrupt controller (GIC)—for managing device interrupts, (ii) DMA controllers (DMAC)—for direct memory access by peripheral devices, and (iii) TrustZone™ Address Space Controller (TZASC)

A. Vasudevan et al., *Trustworthy Execution on Mobile Devices*, SpringerBriefs
in Computer Science, DOI: 10.1007/978-1-4614-8190-4_4, © The Author(s) 2014

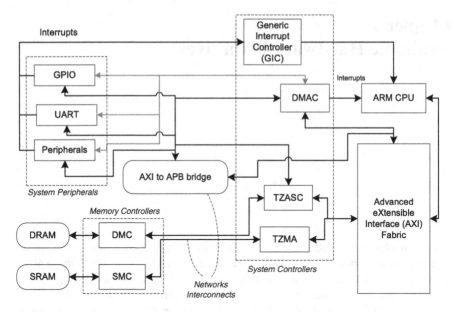

Fig. 4.1 Generic ARM platform hardware and security architecture

and TrustZone™ Memory Adapter (TZMA)—for partitioning memory between multiple "worlds" in a split-world architecture (Sect. 4.2.1).

- *System peripherals* include LCDs, timers, UARTs, GPIO pins, etc. These peripherals can be further assigned to specific "worlds").

We now proceed to discuss the above components in the context of each of the security features described in Chap. 3.

4.2 Isolated Execution

Multiple hardware primitives exist for isolated execution on ARM architecture devices today. ARM first introduced their TrustZone™ Security Extensions in 2003 [7], enabling a "two-world" model, whereby both secure and non-secure software can coexist on the same processor. Today, TrustZone™ features are available for many system components beyond just the CPU(s), as we discuss below.

ARM recently announced hardware support for virtualization for their Cortex™ A15 CPU family [12]. These extensions enable more traditional virtualization solutions in the form of hypervisors or virtual machine monitors [44].

4.2.1 Split-World-Based Isolated Execution

ARM's TrustZone™ Security Extensions [8] enable a single physical processor core to safely and efficiently execute code ins two "worlds"—the *secure world* for

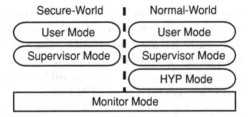

Fig. 4.2 ARM isolated execution hardware primitives. Split-world-based isolation enables both secure and normal processor worlds. Virtualization-based isolation adds a higher-privileged layer for a hypervisor in the normal world

security sensitive application code and the *normal world* for non-secure applications (Fig. 4.2). CPU state is banked between both worlds; the secure-world can access all normal-world state, but not vice-versa. A new processor mode, called the *monitor mode*, supports context switching between the secure-world and the normal-world and can be entered either asynchronously (e.g., as a result of hardware interrupts or exceptions) or synchronously by the execution of the Secure Monitor Call (SMC) instruction. Note that the SMC instruction can only be executed from the supervisor mode (SVC) in the normal-world. The monitor mode software is responsible for context-switching CPU state that is not automatically banked.

4.2.1.1 Memory Isolation

ARM's TrustZone™ Security Extensions split CPU state into two distinct worlds, but they alone cannot partition memory between the two worlds. Memory isolation is achieved using a combination of TrustZone™-aware Memory Management Units (MMU), TrustZone™ Address Space Controllers (TZASC), TrustZone™ Memory Adapters (TZMA), and Tightly Coupled Memory (TCM).

A TrustZone™-aware MMU provides a distinct MMU interface for each processor world, enabling each world to have a local set of virtual-to-physical memory address translation tables. The translation tables have protection mechanisms which prevent the normal-world from accessing secure-world memory. Such MMUs employ tagged Translation Look-aside Buffers (TLB), where entries are tagged with the identity of the world. This enables secure- and normal-world entries to co-exist so as to improve performance [8].

The TZASC interfaces devices such as Dynamic Memory Controllers (DMC) to partition DRAM into distinct memory regions. The TZASC has a secure-world-only programming interface that can be used to designate a given memory region as secure or normal. The TZASC rejects memory transactions from the normal-world that are directed towards secure memory regions. The TZMA provides similar functionality for off-chip ROM or SRAM. With a TZMA, ROM or SRAM can be partitioned between the two worlds.

Tightly Coupled Memory (TCM) is memory that is in the same physical package as the CPU, so that physical tampering with the external pins of an integrated circuit will be ineffective in trying to learn the information stored in TCM. TCMs are typically blocks of fast on-chip SRAM that exist at the same level as the CPU's L1 cache subsystem. Secure-world software is responsible for configuring access permissions (secure versus normal) for a given TCM block.

4.2.1.2 Peripheral Isolation

Peripherals in the ARM platform architecture can be designated as *secure* or *normal*. Secure peripherals are intended to be accessible by the secure world while normal peripherals can be accessed from both worlds. Thus, there is a need to isolate secure and normal peripherals so that software running in the normal world cannot maliciously or inadvertently address secure-world peripherals.

ARM's "CoreLink" architecture connects high-speed system devices such as the CPU and memory controllers using the Advanced eXtensible Interface (AXI) bus [10]. The rest of the system peripherals are typically connected using the Advanced Peripheral Bus (APB). The AXI-to-APB bridge device is responsible for interfacing the APB interconnects with the AXI fabric. The AXI bus transaction packets include an identification field that designates the transaction as secure or normal. However, the APB transactions do not have such a provision [11]. This places the responsibility for managing security-relevant state with the AXI-to-APB bridge.

A TrustZone™-aware AXI-to-APB bridge contains address decode logic that selects the desired peripheral based on the security state of the incoming AXI transaction; the bridge rejects normal-world transactions to peripherals designated to be used by the secure-world. A TrustZone™ AXI-to-APB bridge can include an optional software programming interface that allows dynamic switching of the security state of a given peripheral. This can be used for sharing a peripheral between both the secure and normal worlds.

4.2.1.3 DMA Protection

Certain peripherals (e.g., LCD controllers and storage controllers) can transfer data to and from memory using Direct Memory Access (DMA), which is not access-controlled by the AXI-to-APB bridge. A TrustZone™-aware DMA controller (DMAC) supports concurrent secure and normal peripheral DMA accesses, each with independent interrupt events. Together with the TZASC, TZMA, GIC, and the AXI-to-APB bridge, the DMAC can prevent a peripheral assigned to the normal-world from performing a DMA transfer to or from secure-world memory regions.

4.2.1.4   Hardware Interrupt Isolation

As peripherals can be assigned to either the secure or normal world, there is a need to provide basic interrupt isolation so that interrupts from secure peripherals are always handled in secure world.

Hardware interrupts on the current ARM platforms can be categorized into: IRQ (normal interrupt request) and FIQ (fast interrupt request). The Generic Interrupt Controller (GIC) can configure interrupt lines as secure or normal and enables secure-world software (in monitor mode) to selectively trap such system hardware interrupts. This enables flexible interrupt partitioning models. For example, IRQs can be assigned for normal-world operations and FIQs for secure-world operations. The CPU core provides support for interrupt identification and redirection. For example, if an IRQ occurs during normal-world execution, it is handed over to the normal-world interrupt handler immediately. However, if an IRQ occurs during secure-world execution, the monitor-mode handler is invoked which can choose to handle the IRQ or inject it back to the normal-world. The GIC hardware also includes logic to prevent normal-world software from modifying secure interrupt line configurations. Thus, secure world code and data can be protected from potentially malicious normal-world interrupt handlers, but TrustZone™ by itself is not sufficient to implement device virtualization.

4.2.2   Virtualization-based Isolated Execution

ARM's Virtualization Extensions provide hardware virtualization support to normal-world software starting with the Cortex™ A15 CPU family [12]. The basic model for a virtualized system involves a hypervisor, that runs in a new normal-world mode called Hyp mode (Fig. 4.2). The hypervisor is responsible for multiplexing guest OSes, which run in the normal world's traditional OS and user modes. Note that software using the secure world is unchanged by this model, as the hypervisor has no access to secure world state. The hypervisor can optionally trap any calls from a guest OS to the secure world.

4.2.2.1   ARM Cortex™-A15

The ARM Cortex™-A15 processor architecture includes new capabilities for hardware support for virtualization. The A15 processor retains full compatibility with its predecessors (e.g., Cortex™-A9) and is based on the ARMv7 architecture. It includes hardware support for virtualization extensions which can run multiple OS binary instances simultaneously thereby enabling isolation of multiple execution environments and associated data. The Cortex™-A15 also includes support for multicore processing and Large Physical Address Extensions (LPAE) which provides the ability to use upto 1TB of physical memory. LPAE introduces 40-bit physical

addressing that reduces address-map congestion by providing common global physical addressing while supporting multiple resident virtualized operating systems.

4.2.2.2 ARMv7 Virtualization Extensions

The ARMv7 Virtualization Extensions are similar to the x86 counterpart in terms of
the high-level isolation and virtualization mechanisms. A new non-secure level of
privilege level, called the *HYP* mode holds the hypervisor. The ARMv7 hardware
virtualization extensions provide various mechanisms for the guest such as interrupt
masking, page table management and communication with system interrupt controllers (e.g., GIC) which avoid the need for hypervisor intervention during guest
execution. It also provides configurable traps into HYP mode for various system
control register accesses and instructions. The architecture also provides hypervisor
support for guest instruction emulation via general constructs called *syndromes*. The
ARMv7 virtualization extensions and the HYP mode are designed to co-exist with
the TrustZone™ secure execution architecture as shown in Fig. 4.2.

Two-Level Memory Virtualization

Before virtualization the OS owns the memory and allocates areas of memory to the
different applications. Modern OSes commonly use virtual memory for address space
seperation. With two-level memory virtualization, the address translation is divided
into two stages. Stage 1 translation is owned by each guest OS and Stage 2 translation
is owned by the hypervisor. Tables from Guest OS translate Virtual Address (VA) to
Intermediate Physical Address (IPA) and a second set of tables from the hypervisor
translate the IPA to the final physical address (PA). The hardware allows aborts to
be routed to the appropriate software layer (guest or hypervisor).

Interrupt Virtualization

An Interrupt might need to be routed to one of current or different guest operating
system, the hypervisor or an OS running in the secure TrustZone™ environment.
In the basic model of the ARM virtualization extensions, physical interrupts are
taken initially in the hypervisor. If the interrupt should go to a particular guest, the
hypervisor maps a "virtual" interrupt for that guest.

Virtualization Extensions provide the necessary infrastructure to aid in interrupt
virtualization. More specifically there are special system registers and flag-bits that
are banked (e.g., CPSR.I,A and F bits) which allow a particular guest to change these
bits without the hypervisor needing to trap and emulate them. All virtual interrupts
are routed to non-secure interrupt handlers in HYP mode (e.g., IRQ, FIQ and Aborts).
Finally, the guest manipulates a virtualized interrupt controller while the physical
interrupt controller is in control of the hypervisor.

Device Virtualization Support and DMA Protection

ARM I/O handling uses memory mapped devices. Reads and Writes to the device registers have specific side-effects. Creating virtual devices requires emulation. Typically reads/writes to devices have to trap to the hypervisor which then interprets the operation and performs emulation. Perfect virtualization means all possible devices loads/stores emulated. Unfortunately, fetching and interpreting emulated load/store is performance intensive. ARMv7 hardware virtualized architecture introduces the "syndrome" construct to ameliorate this situation. Essentially, syndromes store information on aborts for some loads/stores. It unpacks key information about the instruction Source/Destination register, Size of data transfer, Size of the instruction, SignExtension etc. which the hypervisor can readily use for emulation purposes.

Providing address translation for devices is an important aspect of any virtualization architecture since it allows containment of device memory accesses in order to enforce isolation. It also allows for unmodified device drivers in the guest OS. If the device can access memory, the guest will program it in the IPA.

ARMv7 hardware virtualization adds the option for a "system MMU" which enables second stage memory translations in the system for devices. A system MMU could also provide stage 1 translations allowing devices to be programmed into the guests VA space. ARM is currently defining a common programming model where the intent is for the system MMU to be present at the system bus level and configurable by the hypervisor.

4.3 Secure Storage

Current ARM platform specifications do not include a root of trust for long-term secure storage. Platform hardware vendors are free to choose and implement a proprietary mechanism if desired. In this section we discuss hardware roots of trust for secure storage that are available on devices today.

4.3.1 Secure Elements

The Secure Element (SE) provides a solution for establishing a root of trust for mobile devices. SEs provide storage and processing of digital credentials and sensitive data in a physically separate protected module such as a smart-card, thereby reducing the physical attack surface. Ideally, the SE provides a flexible secure platform that supports many applications, each of which can be customized and managed independently [15]. Secure elements fall into three broad categories: software SEs, embedded hardware SEs, and removable hardware SEs [45]. For the purposes of this discussion we only consider hardware-based SE solutions.

An embedded SE is an IC fixed to a mobile device to provide a high degree of security for applications handling sensitive data. Embedded SEs are commonly used to provide security for near field communication (NFC) applications such as automated access control, ticketing, and mobile payment systems. For example, Google Wallet™ uses embedded secure elements to store and manage encrypted payment card credentials,[1] so that they are never available to a compromised mobile device OS. Development platforms such as the FreeScale i.MX53 (Sect. 4.8.2) and Texas Instruments M-Shield™ (Sect. 4.8.3), employ an embedded SE to provide a tamper-resistant secure execution and storage environment.

Removable SEs are interfaced to removable memory such as a Secure Digital (SD) Card or Universal Integrated Circuit Card (UICC). With removable SEs, third-party developers can develop applications against a single platform-independent interface. However, removable SEs are readily physically separated from the mobile device (e.g., the SE may be independently lost or stolen). Giesecke & Devrient and Tyfone are notable vendors currently selling removable SEs.

4.4 Remote Attestation

A remote attestation primitive relies on a private key that is exclusively accessible by a small TCB, and the presence of one or more registers to store measurements (cryptographic hashes) of the loaded code (Sect. 3.3). A vast majority of off-the-shelf mobile devices include support for secure or authenticated boot. The boot-ROM is a small immutable piece of code which has access to a public key (or its hash) and authenticates boot components that are signed by the device authority's private key. Platforms such as the FreeScale i.MX53 (Sect. 4.8.2) and Texas Instruments' M-Shield™ (Sect. 4.8.3) contain secure on-chip keys (implemented using e-fuses) that are one-time-programmable keys accessible only from inside a designated secure environment for such authentication purposes. However, none of the hardware platforms, to the best of our knowledge, support platform registers to accumulate measurements of the loaded code. In principle, this support could be added in software by leveraging the hardware isolation primitives and secure storage described previously.

4.5 Secure Provisioning

Current mobile platforms implement mechanisms to authenticate external information, with the hash of the public portion of the signing key stored immutably on the device [33]. However, such capabilities are currently restricted to OEMs or carriers (e.g., software updates, assigning different identities to the device) and remain unavailable for use by arbitrary third-party developers.

---

[1] http://www.google.com/wallet/faq.html

4.6 Trusted Path

Platforms such as M-Shield™ (Sect. 4.8.3) provide basic hardware primitives to realize a trusted path. A special chip interconnect allows peripheral and memory accesses only by the designated secure environment, and secure DMA channels to guarantee data confidentiality from origin to destination. Such capabilities are being used for DRM (video streaming) on certain off-the-shelf mobile devices [27], but it remains unclear if they are available to third-party developers.

4.7 Design Gaps and Challenges

Having described the ARM hardware platform and security architecture and how the different components interplay to provide various hardware security features, we now identify design gaps and implementation challenges in off-the-shelf mobile devices that prevent third-party application developers from fully realizing the desired security features.

ARM's hardware platform architecture is only a specification, leaving the OEMs free to customize a specific implementation to suit their business needs. This means that OEMs could leave out components whose absence can severely constrain some security features and in some cases even break feature correctness. For example, the absence of a TZASC (and/or TZMA) leaves main memory (DRAM/SRAM) accessible to both the secure and normal worlds. The only way to enforce memory isolation between the worlds is to use TCM (Sect. 4.2.1), which has a very limited size (typically 16–32 KB). Similarly, DMA protection requires a TrustZone™-aware DMA controller, GIC, TZASC (and/or TZMA), and a TrustZone™-aware AXI-to-APB bridge. The absence of one of these components will result in the DMA protection being ineffective.

Unfortunately, most of today's off-the-shelf mobile devices include a single set of devices shared between the secure and normal worlds and do not include all the required components to fully realize the hardware security primitives described previously. This results in a huge gap between functional specification and device implementation. OEMs and carriers are generally not concerned with DMA-style attacks or including a TZASC (and/or TZMA) because their physical security requirements already force them to process sensitive data in TCM or other device-specific isolated environments unreachable via DMA.

Many OEMs explicitly lock-out platform security features. For example, TrustZone™ secure-world is enabled or disabled by a single bit in the system configuration register [8]. Once this bit is set to 1 (disabling secure-world), it can no longer be cleared until a device reset. In many off-the-shelf mobile devices such as the Motorola™ Droid™, Droid™-X, BeagleBoard, and some Gumstix platforms, this bit is set to 1 by the boot-ROM code, in essence allowing only normal-world operations.

From a developer's perspective, an abundance of documentation and open-source (or low-cost) development tools are two key factors that facilitate device and platform adoption. While ARM offers decent documentation and development tools (FastModel/RVDS/RTSM) to leverage the hardware security primitives, the cost of the tools (outside of academia) is greater than cost of a typical device. We believe this to be a significant reason why the open-source and hobbyist community has not rallied around ARM's tools.

4.8 Platform Case Studies

We now describe several readily available, inexpensive development platforms that come with a host of interesting security features. These examples serve to show that there is no shortage of *security potential* in mobile device platforms.

4.8.1 ARM Versatile Express

ARM provides a range of both hardware and software development platforms encapsulating both Split-world (TrustZone™) and Virtualization-based (VEAS) isolated execution environments.

The Versatile Express family of development platforms from ARM provides the required environment for prototyping the next generation of system-on-chip designs. These development platforms feature flexible, modular architecture, and high-speed interfaces, hardware and software applications can be developed and debugged quickly and efficiently.

Versatile Express is based on Advanced Microcontroller Bus Architecture (AMBA) and uses the Advanced eXtensible Interface (AXI) or custom logic for use with ARM processors. The Versatile Express system has the following possible sets of component boards (see Fig. 4.3):

- A Motherboard Express (e.g., V2M-P1)
- A CoreTile Express processor daughterboard.
- A LogicTile Express FPGA daughterboard.

The Motherboard Express is especially designed to support latest generations of ARM processors (e.g., with hardware virtualization). It includes all the necessary peripherals that enables porting and developing operating systems and applications for new processors and graphics engines. The motherboard also supports hardware and software application development and debugging through a range of plug-in daughterboards.

The CoreTile Express daughterboard is used together with the Motherboard Express for evaluation, benchmarking and prototyping with the Cortex™-A15 (with hardware virtualization extensions) or Cortex™-A9 through Cortex™-A5 processors (Fig. 4.4).

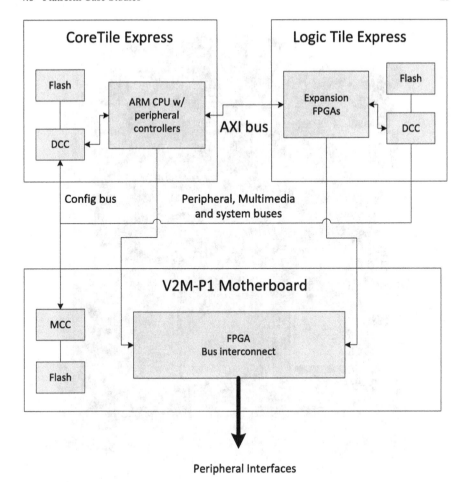

Peripheral Interfaces

Fig. 4.3 ARM Versatile Express Platform Architecture: comprises of a Motherboard Express, a CoreTile Express Cortex™-A series processor daughterboard and a LogicTile Express daughterboard for peripheral connectivity

The Versatile Express LogicTiles provide custom logic expansion capability for the Versatile Express system. They enable peripheral prototyping, validation and software device driver development alongside an ARM processor. A choice of boards with different sized FPGAs and stacking capabilities are available to match design and development requirements.

ARM also provides the RealView Development Suite (RVDS) as a co-ordinated development environment for mobile and embedded systems applications running on the ARM family of processors. RVDS consists of a suite of tools, together with supporting documentation and examples. The tools enable you to write, build, and debug your applications, either on target hardware or software simulators.

Fig. 4.4 ARM Versatile Express Development Kit: supports both TrustZone™ (split-world) and VEAS (Virtualization-based) isolated execution environments based on the Cortex™-A series processors

The ARM software simulators enable development and debugging of software without the requirement for actual hardware. The ARM FastModels and Real-Time System Models (RTSM) provide a Programmer's View (PV) models of processors and devices. The functional behavior of a model is equivalent to real hardware. Absolute timing accuracy is sacrificed to achieve fast simulated execution speed. This means that you can use the PV models for confirming software functionality, but you must not rely on the accuracy of cycle counts or low-level component interactions.

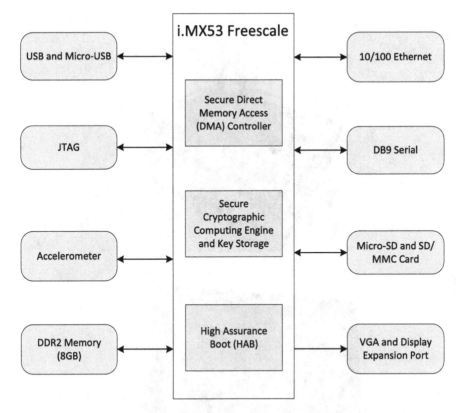

Fig. 4.5 FreeScale i.MX53 Platform Architecture: salient security features include an internal memory that is partitioned into secure and non-secure areas and other features such as secure-boot, secure key storage, and secure DMA controller

4.8.2 FreeScale i.MX53

The FreeScale i.MX53 is a $149 MSRP development board with an ARM Cortex[TM] A8 CPU supporting the Split-world (TrustZone[TM]) based isolated execution environment. Figure 4.5 shows the overall architecture of the i.MX53. The memory subsystem consists of a L1/L2 cache and a L2 internal memory. The L2 internal memory is further divided into a boot-ROM and RAM that is partitioned into secure and non-secure areas. The external memory interface can support upto 2GB DDR3 memory. The i.MX53 makes use of dedicated hardware accelerators to achieve high performance and low power consumption while freeing up the CPU core for other tasks. The system also supports efficient and smart power control and clocking. The i.MX53 is available both as a development board and a tablet form-factor (Fig. 4.6) which makes it a great choice for system prototyping.

The i.MX53 supplements a range of security features that can be used individually or in concert to underpin the platform security architecture. Most of the

Fig. 4.6 FreeScale i.MX53 Development Kit: includes a development board and a tablet form-factor, both running the Cortex™-A8 ARM processor with TrustZone™ (split-world) isolation enabled

i.MX53 security features provide protection against particular kinds of attack and can be configured at various levels according to the required degree of protection. These features are designed to work together and can be integrated with appropriate software to create defensive layers. In addition to protection features, the i.MX53 includes a general-purpose accelerator to enhance the performance of selected industry-standard cryptographic algorithms.

Enhanced TrustZone™

i.MX53 supports the TrustZone™ Architecture (see previously) to provide a trusted execution environment for security-critical software. In addition, the i.MX53 includes a TrustZone™ Interrupt Controller (TZIC) and a TrustZone™ Watchdog (TZWDOG). The TZIC collects interrupt requests from all sources and provides the interrupt interface to the core. Each interrupt source can be configured dynamically as a Normal or a Secure interrupt.

The TrustZone™ Watchdog (TZWDOG) protects against Normal World software preventing a switch back to the Secure World, thereby starving security services of access to the core. Once the TZWDOG is activated, it must be serviced by Secure World software on a periodic basis. If servicing does not take place before the configured time-out, the TZWDOG asserts a secure interrupt that forces a switch to the Secure World. If it is still not served, the TZWDOG asserts a security violation alarm. The TZWDOG cannot be programmed or deactivated from the Normal World.

Secure-Boot Process

Unauthorized software can enter the platform during upgrades or re-provisioning, or when booting from USB/UART connections or removable devices. If permitted to gain control of the boot sequence, unauthorized software can be the attack vector for a variety of goals including exposing stored secrets, circumventing access controls to sensitive data, services or networks, and re-purposing the platform. The i.MX53 supports a High Assurance Boot (HAB) process where the system boot-ROM prevents the platform from executing unauthorized software during the boot sequence. Using digital signatures to recognize authentic software, HAB supports booting the device to a known initial state, running software signed by the (lifetime-write-once) designated authority.

Secure Cryptographic Key Storage

The i.MX53 Security Controller provides a small Secure RAM area that is self-clearing on tamper detection or software deallocation. The code can execute out of the Secure RAM. The Secure RAM is divided into 4 seperate areas called partitions. The security controller is TrustZone™-aware and provides configurable access controls for each partition.

Secure Cryptographic Computing Engine

The i.MX53 Security Accelerator (SAHARA) provides a dedicated cryptographic engine for importing data to or exporting data from Secure RAM. It has a 256-bit dedicated secret master key that is protected from other software or hardware accesses.

Individual partition keys are bound to partition allocation, permissions and software-supplied values. The SAHARA has a dedicated TrustZone™-aware DMA controller and accelerates the following cryptographic functions: AES, DES/3DES, ARC4, MD5, HMAC, SHA-1, SHA-224 and SHA-256. It also features entropy generation.

DMA Controller

The i.MX53 Smart Direct Memory Access (SDMA) controller is a software programmable DMA controller that enables data transfers between peripheral I/O devices and internal/external memories. The SDMA supports two security levels: (a) open mode—where the CPU has full control to load scripts and execution context into SDMA RAM and modify SDMA registers; and (b) locked mode—where selected SDMA registers become read-only to prevent modification of software reset, exception, and debug handling.

4.8.3 Texas Instruments OMAP™ and M-Shield™

OMAP™ (Open Multimedia Applications Platform) developed by Texas Instruments is a category of proprietary system on chips (SoCs) for portable and mobile multimedia applications. OMAP™ devices include a general-purpose ARM architecture processor core plus one or more specialized co-processors. OMAP™ SoCs are found in many mobile phones including those from Nokia™ (N series), Motorola™ (Droid™) and Blackberry™.

The OMAP™ family consists of three processor groups classified by performance and intended application: (a) High-performance applications processors, (b) Basic multimedia applications processors, and (c) Integrated modem and applications processors. The High-performance applications processors are chiefly used in smartphones today with processors powerful enough to run significant operating systems (such as Linux™, Android™ or Symbian), support connectivity to personal computers, and support various audio and video applications. The Basic multimedia and Integrated modem processors are intended to be highly integrated for use in low cost cell phones.

Figure 4.7 shows the OMAP™ 4 platform architecture. We briefly discuss the salient OMAP™ technologies below:

- **Connectivity and System Integration** The OMAP™ platform provides multiple wireless connectivities such as Bluetooth, Wi-Fi, GPS and FM. Beyond wireless connectivity OMAP™ platforms also include pre-integration of a number of application-specific protocols such as SD, EMMC, Ethernet, USB, SATA and PCI Express.
- **Programmable DSP** OMAP™ processors contain a programmable DSP, which can accelerate the decoding of images in real time. While other processors rely on

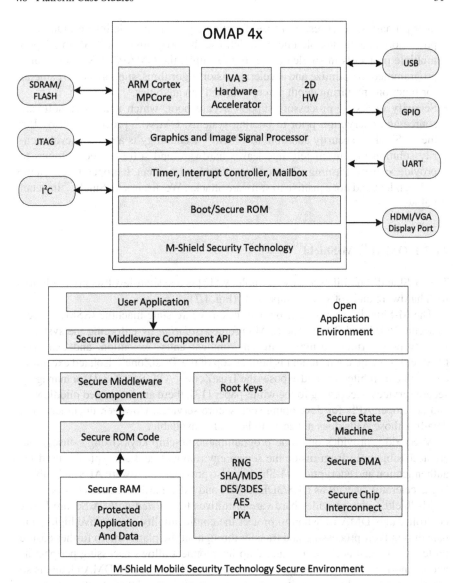

Fig. 4.7 OMAP™ (Open Multimedia Applications Platform) Architecture: The OMAP™ platform provides multiple connectivity options (Wi-Fi, SD, Ethernet, USB etc.) and also contains a programmable DSP which can be used to optimize image decoding and run various analytics. OMAP™ platforms offer the M-Shield™ system-level security solution that provides a trusted execution environment with other security features such as secure-boot, secure DMA and public-key infrastructure

their primary CPU cores, or a fixed function engine to decode images, OMAP™ processors enable flexible clients which use the programmable DSP to support multiple protocols in single device. For example, the OMAP™ DSP can be also programmed to optimize and accelerate vision algorithms, support stereo cameras for functions requiring depth perception, and run analytics.

- **Security** OMAP™ processors support secure boot, which authorizes software source and encryption prior to booting systems. OMAP™ processors also offer the M-Shield™ security solution. M-Shield™ technology is a system-level security solution that intimately interleaves hardware and software technologies to provide security, forming a Trusted Execution Environment. It is operating system-independent and not sensitive to software attacks. We discuss M-Shield™ in detail below.

4.8.3.1 OMAP™ M-Shield™

The M-Shield™ mobile security technology [13] is a system-level security solution with hardware and software components (Fig. 4.7).

The M-Shield™ secure environment has a secure state machine (SSM) as well as secure ROM and RAM. The SSM enforces *isolation* by enforcing the system's security policy rules during secure environment entry, execution, and exit. The M-Shield™ secure environment is built on top of the TrustZone™ isolated execution environment architecture and exposes the TrustZone™ API (Sect. 6.3.3) for managing secure services. According to the white-paper [13], there are associated middleware and developer APIs for developing such secure services. However, documentation detailing those APIs does not seem to be readily available.

M-Shield™ provides one-time programmable on-chip keys (using e-fuses) that are accessible only from inside the secure environment, and are typically used for authentication and encryption. M-Shield™ also provides a hardware AES and public-key accelerator, as well as DES/3DES, SHA and MD5 hardware accelerators.

M-Shield™ also provides hardware primitives for *trusted path*. A Secure DMA controller tags DMA transfers to protect the confidentiality of sensitive high-value data during their processing and transfer throughout the platform. To further ensure protection against attacks, a secure chip-interconnect allows accessing peripherals and memories only by the secure environment and/or by secure DMA channels so that the confidentiality of sensitive information is guaranteed through the entire data path, from origin to destination.

M-Shield™ also includes a public-key infrastructure that provides a secure means to validate the authenticity and integrity of software on the platform before execution, thereby supporting *secure-boot* and enabling *authenticated-boot*.

4.8.3.2 OMAP™ Development Boards

OMAP™ (M-Shield™) SoCs are used as the basis for a number of hobbyist and prototyping boards. We examine a few popular boards and their support for developing security sensitive applications.

The Beagle Board, Panda Board and Gumstix are low-cost, fan-less single-board computers based on the OMAP™ 3 device family, with all of the expandability of today's desktop machines, but without the bulk, expense, or noise (Fig. 4.8). At the heart is the ARM Cortex™-A8 processor with TrustZone™ support. The design goal of these boards was to make it as simple and cheap (they boards retail less than $150) as possible e.g. not having a LCD added, but letting you connect all add-ons available as cheap external components. They are reported to run Linux™ and Android™ OSes.

Logic's Zoom Mobile Development Kit is a form-factor development platform that leverages the processing capabilities of OMAP™ while offering developers a more realistic system for development and validation. The platform runs a Cortex™-A8 ARM processor and can boot Linux. The kit comes with various software resources including supporting kernel and libraries.

Unfortunately, all the above development boards have TrustZone™ disabled [2] and lack access to any of the hardware security features.

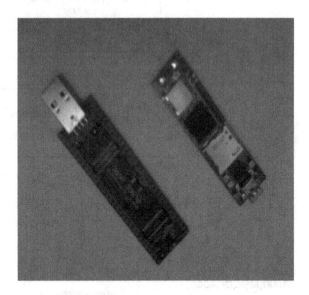

Fig. 4.8 OMAP™ development boards such as the Gumstix are readily available low-cost fan-less single-board computers based on the OMAP™ platform. Other OMAP™ development boards include the Beagle Board, Panda Board and Logic's Zoom Mobile Development Kit. Unfortunately, all the above development boards disable TrustZone™ within the boot-ROM during power-up and lack access to any of the security features specified by the OMAP™ M-Shield™

---

[2] The Initial Program Load (IPL) ROM code seems to switch the Cortex™-A8 processor into Normal world immediately on boot-up. http://e2e.ti.com/support/omap/f/849/t/58680.aspx.

4.8.4 Samsung Exynos^{TM}

The Exynos™ 5 is System-on-a-chip (SoC) by Samsung that is based on the ARM Cortex™-A15 processor. The Exynos™ 5 SoC provides a dual-core CPU, WQXGA display and specialized 2D/3D graphics hardware. The platform also includes a dedicated Image Signal Processor and connectivity to high-speed peripheral interfaces such as USB 3.0 and SATA3.

Figure 4.9 shows the high-level architecture of the Exynos™ 5 platform. At its core, the platform includes an ARM Cortex™-A15 dual-core processor. Exynos™ 5 also features in addition the quad-core ARM ARM Mali T604 GPU to deliver superior GPU performance. The advent of GPGPU's, which represents a substantial change

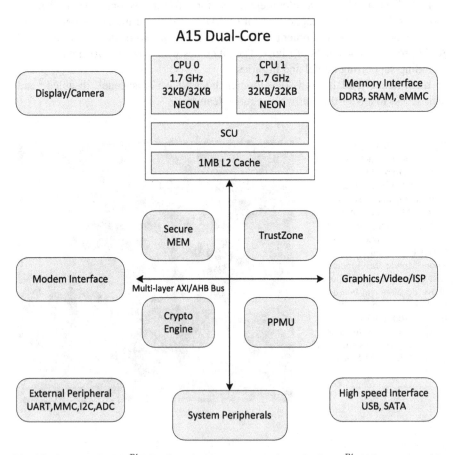

Fig. 4.9 Samsung Exynos™ 5 Platform Architecture: comprises of a Cortex™-A15 processor with support for both TrustZone™ and Virtualization-based isolated execution environments. The platform also features a GPGPU for MIMD parallelism and supports various other security features such as secure-boot, hardware cryptographic accelerators and key management

in hardware architecture enables massive MIMD parallelism and utilize bandwidth to get the most out of Exynos™'s shared memory architecture.

The Exynos™ 5 also incorporates several security features. The platform incorporates the TrustZone™ (split-world) isolated execution architecture with the ability to run secure sensitive applications in a secure world. The Cortex™-A15 hardware virtualization extensions enable the HYP mode or the hypervisor mode that can also be used to implement an isolated execution environment to enforce desired security properties.

The security subsystem also supports hardware cryptographic accelerators for AES, DES/3DES, ARC4, SHA-1/SHA-256/MD5/HMAC/PRNG and TRNG. The platform also supports secure-boot and contains a dedicated secure RAM that is only accessible to the TrustZone™ secure-world. The platform also features a run-time integrity check subsystem which can be configured to check memory data integrity during runtime.

4.8.4.1 Exynos™ Development Kit

The Arndale Board, a new community development board is designed around the Exynos™ 5 Dual system-on-chip (SoC). The development board offers the open source developer community a rich environment for producing the highest caliber of mobile applications focused in the areas of security, gaming, multimedia, and user interface on multiple operating systems. The Arndale board also includes features such as Near Field Communication (NFC), a Global Positioning System (GPS) and a camera sensor. Wi-Fi and Bluetooth connectivity are also provided. The development kit also features an add-on LCD component that can convert the development board into a tablet-form factor.

Chapter 5
Isolated Execution Environments

An execution environment that is isolated from the device operating system (Sect. 3.1) is perhaps the most critical security feature described in Chap. 3. Such an environment can be used to run secure services that multiplex hardware-backed security features, such as secure storage (Sect. 3.2), amongst the various stake-holders, including third party application developers.

Greater flexibility can be offered to third-party developers by allowing them to run modules inside that environment. This mechanism provides the strongest security for those modules, since data can be prevented from leaving the secure environment. However, it also requires ensuring that software modules in that environment cannot compromise each other, the environment itself, or the main OS. While this increases the size and complexity of the isolated environment's trusted-computing-base, such an environment remains smaller and more trustworthy than a full-featured OS.

The available isolated-execution hardware primitives Sect. 4.2 offer several options for implementing isolated execution environments. We consider two high-level approaches: either using a parallel execution environment, or multiplexing a single execution environment using a hypervisor.

5.1 Parallel Isolated Execution

One strategy for isolated execution is to put sensitive code in a distinct, parallel environment. As described in sect. 4.2.1, current ARM platforms that support TrustZone™ offer a mechanism by which secure software can execute in isolation within a special processor world. Several research proposals [21–23, 34, 62, 65] employ TrustZone™ to achieve isolation and provide a subset of the security properties discussed in Chap. 3. Other approaches make use of a physically separate protected module such as a smart-card to achieve isolation. One notable example is the Trusted Execution Module (TEM) [17], which is capable of securely executing procedures (called *closures*) expressing arbitrary computation. The TEM itself is a

A. Vasudevan et al., *Trustworthy Execution on Mobile Devices*, SpringerBriefs
in Computer Science, DOI: 10.1007/978-1-4614-8190-4_5, © The Author(s) 2014

byte-code interpreter for a small special-purpose programming language. This interpreter is realized as a JavaCard applet, hosted inside a JavaCard-enabled [51] smartcard. Another example is a smart-card-based Mobile Trusted Module (MTM) [20] that implements the MTM functionality in Java applets that can be downloaded into the smart-card. We provide a detailed discussion of the above frameworks in Chap. 6.

5.2 Hypervisors

A *microkernel* is a minimal OS, with many components that would be part of a monolithic OS, particularly device drivers, either removed or running as deprivileged processes. A *hypervisor* is a microkernel that can run other OSes as deprivileged processes. OSes can run unmodified if the environment provided by the hypervisor (optionally with help from some of its deprivileged services) matches the physical hardware expected by that OS. Otherwise we say that the OS must be *para-virtualized*—modified to run in the environment that is provided by the hypervisor.

A hypervisor can be used to implement an execution environment that is isolated from the main OS by running the operating system as one process (a virtual machine), and by running the modules to-be-isolated as separate processes.

5.2.1 Hypervisor Attributes for Mobile Devices

Hypervisors on mobile devices implement a different kind of abstraction with different constraints than other platforms. This section explores some of the constraints and capabilities provided in the mobile device space.

Although mobile devices have been commonly associated with severe resource constraints, today's devices range from powerful processors with server-level functionality (such as hardware support for virtualization) to power-optimized systems with less compute capacity and resources. This variety creates a more demanding environment for hypervisors on mobile devices than their mainframe and server siblings.

Efficiency

All hypervisors strive for efficiency, but hypervisors on mobile devices must deal with added constraints outside of traditional virtualization environments. Outside of processor sharing, memory tends to be one of the key limiters to performance in mobile device environments. For this reason, hypervisors designed for mobile devices must be small and extremely efficient in their use of memory.

Security

A hypervisor being small has its advantages. The smaller the code size of an application, the easier it is to validate and prove that it is bug free. In fact, some mobile device hypervisor vendors have formally verified their hypervisors and guaranteed them to be bug free (e.g., seL4 [32]). The smaller the hypervisor, the more secure and reliable the platform can be. This is because the hypervisor is typically the only portion of the system to run in a privileged mode, which serves as what is known as the Trusted Computing Base (TCB) and leads to a more secure platform.

Communication

Hypervisors for mobile devices are built for sharing a hardware platform with multiple guests and applications but also commonly extend communication methods to allow them to interact. This channel for communication must be both efficient and secure, permitting privileged and non-privileged applications to coexist.

Isolation

Related to security is the ability to isolate guests and applications from one another. In addition to providing containment for security and reliability, it provides benefits in terms of license segregation. Using the mobile device hypervisor's communication mechanism permits proprietary software and open source software to coexist in isolated environments. As mobile devices become more open, the desire to mix proprietary software with third-party and open source software is a key requirement.

Real-Time Capabilities

Finally, the mobile device hypervisor must support scheduling with real-time capabilities. In the case of mobile phones, the hypervisor can share the platform with core communication capabilities and third-party applications. Scheduling with real-time characteristics allows the critical functions to coexist with applications that operate on a best-effort basis.

5.3 Hypervisor Case Studies

We now describe in detail some noteworthy existing ARM hypervisor projects shown in Table 5.1. Current closed-source hypervisors include Winter [62], seL4 [32], OKL4 [39], and INTEGRITY [28]. Winter outlines an approach to merge TCG-style Trusted Computing concepts with ARM TrustZone™ technology in order to

Table 5.1 Noteworthy ARM hypervisors and microkernels

HyperVisor/Microkernel	Virtualization type	Code availability	Maturity level
Winter	Split-world	Closed-source	Unknown
SeL4	Para-virtualization	Closed-source	Unknown
OKL4	Para-virtualization	Closed-source	Mature
INTEGRITY	Para-virtualization	Closed-source	Mature
KVM/ARM	Full-virtualization	Open-source	Work-in-Progress
CodeZero	Para-virtualization	Open-source	Work-in-Progress
EmbeddedXen	Para-virtualization	Open-source	Work-in-Progress
Xen/ARM	Para-virtualization	Open-source	Work-in-Progress
XMHF	Full-virtualization	Open-source	Work-in-Progress

build an open Linux™ based embedded trusted computing platform. The seL4 project gained notoriety in 2009 when they announced a formally verified microkernel for the ARM architecture. OKL4 is a microkernel-based embedded hypervisor with a small footprint and CPU support to target mobile telephony. The INTEGRITY multivisor uses a security kernel to provide domain isolation and is targeted at in-vehicle infotainment and next-generation mobile devices.

KVM/ARM,[1] Codezero,[2] EmbeddedXen,[3] XenARM [63], and XMHF[4] are some noteworthy open-source hypervisor initiatives. We now describe these hypervisors in more detail.

5.3.1 KVM/ARM

The KVM for ARM project is focused on creating a open-source KVM-based virtualization solution for ARM-based devices that can run virtual machines efficiently. The project explores both software-only and hardware-assisted approaches, the latter by leveraging ARM virtualization extensions. Figure 5.1 shows the overall hypervisor architecture.

The software-only approach uses lightweight paravirtualization, a script-based method to automatically modify the source code of an operating system kernel to allow it to run in a virtual machine. Lightweight paravirtualization is architecture specific, but operating system independent. It is minimally intrusive, automated, and requires no knowledge or understanding of the guest operating system kernel code [19].

---

[1] http://wiki.ncl.cs.columbia.edu/wiki/KVMARM:MainPage

[2] http://www.l4dev.org

[3] http://sourceforge.net/projects/embeddedxen

[4] http://xmhf.org

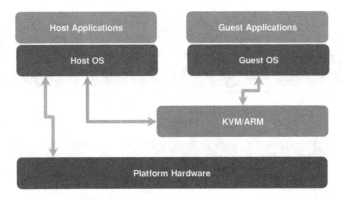

Fig. 5.1 KVM/ARM hypervisor architecture

The basic idea behind lightweight paravirtualization is to modify the guest kernel source-code to replace sensitive non-privileged instructions so that they can trap to the hypervisor. The hypervisor then emulates the instruction. The key idea behind the automation is to use a script based on regular expressions that parses the kernel source code and performs such replacements automatically. The current KVM/ARM software-only prototype is based on the Linux™ kernel used in Google™ Android™, and is reported to run nearly unmodified Linux™ guest operating systems [5].

The KVM/ARM hardware-assisted approach focuses on ARM support for hardware virtualization extensions. More specifically the current KVM/ARM hardware-assisted prototype operates on the Cortex™-A15 processor using the ARM FastModels emulator as well as the ARM Versatile Express hardware platform. At a high level the KVM/ARM hardware-assisted approach leverages the ARM HYP mode to context switch from host to guest and back. On every such context switch, KVM saves and restores host and guest execution contexts (e.g., translation tables, trap configurations, general purpose and system registers etc.). KVM also pre-empts the guest on certain conditions such as interrupt delivery, translation faults and few privileged system register accesses and cache maintenance operations. On guest exit, the control is transferred to the host which then hadles the exit accordingly (e.g., interrupts, page-faults). The hypervisor has no influence at all when running the host. The current KVM/ARM hardware-assisted prototype is reported to successfully run unmodified Linux™ guest operating systems [60].

5.3.2 CodeZero

The Codezero hypervisor is a new microkernel that follows the L4 architecture but has been written from scratch to benefit from the latest research in microkernel design. It follows the fundamental principles of microkernels in that it implements address spaces, thread management, and IPC only in the privileged microkernel along with virtualization capabilities. The current Codezero prototype supports the ARMv7

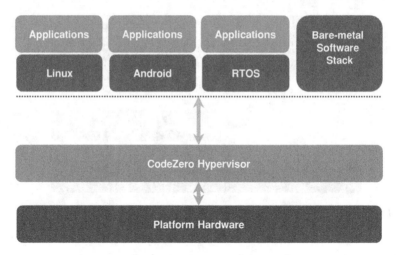

Fig. 5.2 CodeZero hypervisor architecture

(hardware-virtualized) architecture and is reported to run multiple operating systems at the same time on a single platform while ensuring a hardware-based isolation between them.

As shown in Fig. 5.2, Codezero implements a typical abstraction layer over the hardware platform. The abstraction layer implements threading, IPC, address space management, address space mapping, security, power, and error recovery management. Codezero's scheduler includes kernel preemption for both guest threads and microkernel threads (in addition to time slices for preemption).

Virtualization in Codezero is implemented through containers. Each container is an isolated execution environment with its own set of resources (memory, threads, and so on). The partition also works in concert with Codezero's security and resource management policies, which define capabilities for each container. Secure containers can be created as many as needed, with the ease of software. Each secure container is then populated with additional functionality, from simple applications to full blown operating systems such as Android™ or Linux™.

Codezero benefits from recent advancements in microkernel designs. For efficiency, Codezero implements three forms of IPC (all based on the rendezvous model). Codezero implements short IPC (between user space threads), full IPC (256 bytes), and extended IPC (2048 bytes). IPC of larger buffers is performed through shared-page mappings. Codezero was also designed specifically for mobile and embedded platforms and supports multicore processors as well as ARM-based designs.

5.3.3 OKL4

In 2006, Open Kernel Labs (OK Labs) was founded for the development of micro-kernels and hypervisors for mobile and embedded systems. The Lab's work in each of these domains coined the term microvisor, which represents a microkernel with virtualization capabilities. OK Labs is by far the most successful in the space of embedded virtualization, deploying its open source OKL4 microvisor into more than a billion devices, such as the Evoke QA4 messaging phone, the first phone to support virtualization and operation of two concurrent operating systems (including Linux™).

The heritage of OKL4 comes from the L4 family of microkernels. L4 in turn was inspired by Mach (the Carnegie Mellon University microkernel developed as a drop-in replacement for the traditional unix kernel). L4 was originally designed entirely in x86 assembly in order to realize an optimal solution. It then was developed in C++ and exists in a family of microkernels (from L4Ka::Hazelnut, designed for Intel Architecture, 32-bit, and ARM-based architectures, to L4Ka::Pistachio, designed for platform independence and released under the Berkeley Software Distribution license).

The OKL4 microvisor implements partitions known as secure cells for partitioning VMs in the architecture (Fig. 5.3). The OKL4 microvisor occupies the privileged kernel space, and all VMs, native applications, and drivers are pushed into separate isolated partitions with an efficient interprocess communication (IPC) mechanism to allow cells to communicate and cooperate (see Fig. 4). In addition to traditional IPC between VMs, because hardware device drivers are pushed outside of the microvisor (as is typical with microkernels), the IPC is important: It is a common path input/output. Further, because individual applications and drivers can be integrated into the platform without an operating system, the component model for OKL4 is lightweight.

The microvisor implements the core microkernel with virtualization capabilities, which includes resource management as well as scheduling with real-time capabilities and low performance overhead. OKL4 implements paravirtualization, which

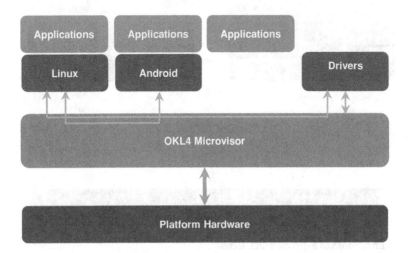

Fig. 5.3 OKL4 hypervisor architecture

means that operating systems must be instrumented to run on the microvisor. OK Labs provides support for a number of paravirtualized operating systems, including OK:Linux, OK:Android, and OK:Symbian.

5.3.4 EmbeddedXen

EmbeddedXEN is a particularly efficient virtualization framework tailored to ARM-based core mobile and embedded systems. While security and OS isolation are key features of conventional virtualization frameworks, the main concerns for EmbeddedXEN are device heterogeneity and realtime aspects.

EmbeddedXEN mainly relies on the original XEN architecture but with major differences in the way guest OS are handled: the hypervisor has been simplified, and only two guest OS (dom0 and domU) can run simultaneously; while dom0 is used to manage the native OS with drivers (original and backend splitted drivers), a paravirtualized OS (domU) can be cross-compiled on a different ARM device, and user applications can run seamlessly on the (virtualized) host device (Fig. 5.4).

Another important difference is that no user space tools are required to manage the VMs; the framework produces a compact single binary image containing both dom0 and domU guests, which can be easily deployed. The Xenbus architecture has been adapted to that context.

EmbeddedXEN therefore allows the porting of an OS and its applications from an ARM embedded device to latest generation ARM hardware, such as HTC Smartphone for example.

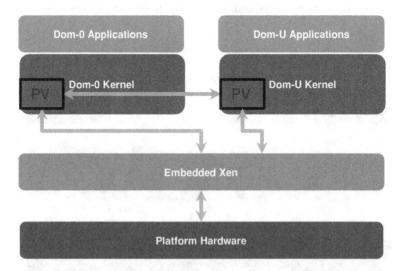

Fig. 5.4 EmbeddedXen hypervisor architecture

5.3.5 *Xen/ARM*

The Xen/ARM project maintains an ARM variant of the Xen Hypervisor in a codeline that is separate from the upstream Xen Hypervisor project. The Xen ARM Project is led by Samsung, and delivers and maintains Xen support for a range of ARM processors (ARM v5–v7) for mobile devices, using Xen Paravirtualization (PV). The project is also working on problems such as solving real-time guarantees in a virtualized environment and multi-processor support.

Xen Paravirtualization (PV) is an efficient and lightweight virtualization technique introduced by Xen, later adopted by other virtualization solutions. Xen PV does not require virtualization extensions from the host CPU and thus enables virtualization on hardware architectures that do not support hardware-assisted virtualization. However, Xen PV guests and control domains require kernel support and drivers that in the past required special kernel builds, but are now part of the Linux™ kernel as well as other operating systems.

Xen PV delivers higher performance than full virtualization because the operating system and hypervisor work together more efficiently, without the overhead imposed by the emulation of the system's resources. This makes a big difference for disk and network operations, where the use of Xen PV network, bus and block device drivers enable near-native performance. Examples of devices which benefit from Xen PV and where drivers are available are block (disks), SCSI, USB, VGA and PCI devices. Architecturally, Xen PV works by opening additional channels of communication between the hypervisor and the guest operating systems via Xen PV front end and back end drivers as shown in Fig. 5.5.

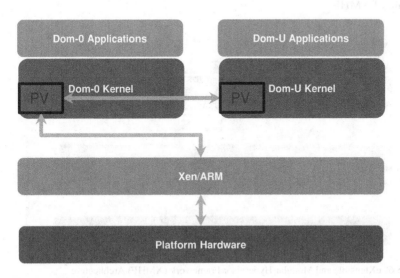

Fig. 5.5 Xen/ARM hypervisor architecture

With the introduction of virtualization extensions on ARM processors, the Xen community has taken steps to add ARM support for ARM CPUs to the Xen Hypervisor. This port is executed as part of of Xen hypervisor project, with no separate codebase.

Xen/ARM with virtualization extensions is re-architected to exploit the (ARM) platform hardware as much as possible. It only supports one type of guest which is a modified Linux™ kernel with Xen PVH paravirtualization. Xen PVH is esentially paravirtualized interfaces for I/O while running in a hardware virtual machine. Xen PVH guests are essentially PV guests using PV drivers for boot and I/O. Otherwise it uses hardware virtualization extensions, without the need for emulation. PVH has the potential to combine the best trade-offs of all Xen virtualization modes, while simplifying the Xen architecture significantly. The current Xen/ARM with hardware virtualization support reportedly runs on currently runs on ARM RSTM/FastModels emulator, ARM Versatile Express platform with Cortex™-A15 processor and work is also underway to support the Samsung Exynos™ platform with the Arndale development board.

5.3.6 eXtensible Modular Hypervisor Framework

The eXtensible and Modular Hypervisor Framework (XMHF)[5] takes a developer-centric approach to hypervisor design and implementation, and strives to be a comprehensible and flexible platform for performing (security-oriented) hypervisor research and development on commodity computing platforms. Figure 5.6 shows the architecture of XMHF.

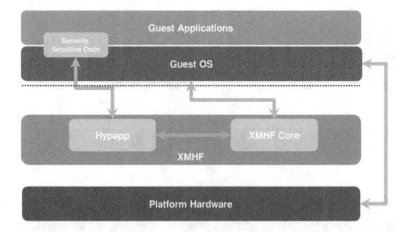

Fig. 5.6 eXtensible and Modular Hypervisor Framework (XMHF) Architecture

---

[5] http://xmhf.org

XMHF encapsulates common hypervisor core functionality in a framework that allows others to build custom hypervisor-based solutions (called "hypapps"). As a small piece of software between the OS and the hardware, hypapps therefore enjoy a unique advantage in terms of balance between security and versatility. They also help reduce developer concerns with respect to other malicious applications within the OS or OS vulnerabilities. Further, the hypapps can be architecture/platform independent while relying on XMHF to provide the necessary platform support.

A significant design decision in XMHF motivated by minimalism is the support of only a single full-featured commodity guest OS (rich guest). XMHF leverages hardware virtualization primitives to allow the guest direct access to all performance critical system devices and device interrupts. This model results in reduced hypervisor complexity (since all devices are directly controlled by the OS) and consequently TCB, while at the same time promising near-native guest performance.

XMHF's minimal hypervisor design also enables automated and development compatible verification of its C code implementation. XMHF has been successfully verified at the source-code level using the CBMC model-checker to guarantee the fundamental hypervisor security property of memory integrity [58]–that the hypervisors memory must not be modified by software running at a lower privilege level. The combination of hardware virtualization primitives, and XMHF design and development choices results in an architecture where manual auditing applies primarily to a very small set of functions, and memory integrity can be automatically re-verified in the face of common development changes.

XMHF currently supports both Intel and AMD x86 hardware virtualized platforms and is capable of running unmodified legacy multiprocessor capable OSes such as Windows and Linux. Work is currently underway to support both ARM TruztZone and Virtualization Extensions with the Android™ OS.

XMHF provides a good starting point for research and development on hypervisors with rigorous and "designed-in" security guarantees. Given XMHF's features and performance characteristics, it has the potential to significantly enhance (security-oriented) hypervisor research and development.

5.4 Discussion

Hypervisor frameworks potentially hold value for all stake-holders (OEMs, carriers developers, and users). From an OEM perspective, secure hypervisor frameworks allow multiplexing security-critical baseband functionality on the same processor as popular unmodified OSes and user-facing applications, thereby reducing the cost of materials in a smartphone [39, 42]. Indeed, this appears to be OK Labs' primary business model. From a developer stand-point, hypervisor frameworks allow creation of custom security applications that can benefit from improved isolation (e.g., mobile banking and payments or anti-malware). From a user's perspective, a hypervisor framework may enable simultaneous execution of different OSes, offering a rich set of security features and execution environments on a single mobile device.

Hypervisors are deployed in custom (OEM- and carrier-specific) environments on roughly 1 billion off-the-shelf mobile devices [39, 42]. These can be, and likely already are, used to run security-critical services in isolation from a fully-featured OS running on the same CPU. Unfortunately, we observe that this is done transparently to the user and to third-party developers. These devices do not provide an open API to third-party developers to run *their own modules* in an isolated execution environment provided by the hypervisor.

5.4.1 Limitations of Paravirtualization

All known ARM hypervisors except for KVM/ARM and XMHF, use paravirtualization to support guest OSes (Table 5.1). While paravirtualization in general has proven successful, and many individual drivers are paravirtualized on many commodity platforms such as x86, there is an unavoidable additional maintenance cost for paravirtualization. Unless the paravirtualized hardware architecture and corresponding OS and driver changes are accepted as a first-class architecture by the OS kernel, the maintainers of the paravirtualization-related changes will perpetually be playing catch-up.

Thus, the price of paravirtualization is increased maintenance cost and more limited availability in terms of supported guest operating system versions. For example, Samsung's Xen for ARM requires modifying approximately 5000 lines of code in the Linux kernel [63]. The most recent kernel version it can support is a modified Linux 2.6.11 kernel, a relatively old version of Linux released in 2005.

Chapter 6
API Architectures

Having discussed the hardware primitives available on today's mobile platforms in Chap. 4 and how those can be used to implement reduced-TCB isolated execution environments in Chap. 5 we now discuss potential application programmer interfaces (APIs) that those isolated execution environments may expose to developers.

6.1 API Types

We distinguish between two types of APIs: *App-IEE* APIs and *Module-IEE* APIs.

App-IEE APIs specify how normal applications running on the main OS interact with the isolated execution environment. Such APIs could include mechanisms for communicating with modules running inside the isolated execution environment, for discovering service-modules running inside the isolated execution environment, or for loading third-party software modules into the isolated execution environment.

Module-IEE APIs specify how to develop modules running inside the isolated execution environment. As discussed in Chaps. 3 and 5 such environments will typically minimize their size and complexity by not offering the functionality of a full-featured OS. Instead, these APIs will typically offer some or all of the security services discussed in Chap. 3 some APIs for communicating with software running on the full OS, and possibly some APIs for communicating directly with peripherals (i.e., a trusted path to those peripherals).

6.2 App-IEE-Only Model Versus App-IEE + Module-IEE Model

A minimal way to make hardware security features available to application developers is for OEMs or network carriers to provide security-relevant services running inside the isolated execution environment, and expose them via App-IEE APIs.

This approach may be attractive to OEMs and carriers, who may not want to bear the risk of allowing third-party code to run in the device's isolated environment, or the cost of implementing strong isolation between modules in that environment. Unfortunately, without providing application developers with an isolated execution environment (Sect. 3.1) in which to run their modules, the security properties gained have a large TCB that includes the entire OS at runtime. Still, even this strategy improves the set of security features available to third-party developers today, as we detail below.

We first summarize the desirable properties that arise when a Module-IEE API for running custom code in the isolated execution environment *is* available to application developers. Module-IEE APIs for secure storage enable developers to ensure that only their module can access sealed data, even if the OS is compromised. Module-IEE APIs for remote attestation can run code isolated from the OS, and need not include the OS's measurements in their remote attestations. Module-IEE APIs for secure provisioning can ensure that only the intended module running in the isolated execution environment will be able to access provisioned data. A trusted path implemented via Module-IEE APIs can provide assurance to the user that he is communicating with the intended module running in the isolated execution environment.

We now summarize the benefits to application developers that arise from OEM- or carrier-provided security services exposed through an App-IEE interface. Secure storage (Sect. 3.2) can be implemented by allowing direct access to a secure storage location, or by implementing a sealed-data API. Data sealed in this way would be protected from offline attacks, and attacks where a different OS is booted (since the sealed-data-service would refuse to unseal for the modified OS). Remote attestation (Sect. 3.3) implemented in the App-IEE-only model can attest that a known OS image booted. This can provide some assurance to remote parties that they are communicating with a client that started in a known configuration. However, such mechanisms cannot detect if the OS has been compromised after it was booted. Similarly, a secure provisioning (Sect. 3.4) service built in the App-IEE-only model can ensure that exported data can only be accessed by a known device that booted a known OS. However, it would have to trust that OS to not compromise the data itself or to allow unauthorized applications to access that data. A trusted-path service (Sect. 3.5) implemented in the App-IEE-only model can ensure to the user that an authorized OS booted, but not that the OS remains uncompromised after it has booted.

6.3 Candidate APIs

We next discuss several published APIs. All of these specify App-IEE APIs; some of them additionally specify Module-IEE APIs.

6.3.1 Mobile Trusted Module

The Mobile Trusted Module (MTM) is a specification by the Trusted Computing Group (TCG) for a set of trusted computing primitives [52]. Like the Trusted Platform Module on PCs, the MTM provides APIs for secure storage and for attestation, but does not by itself provide an isolated execution environment for third-party code or facilities for trusted path.

Unlike the TPM, the MTM is explicitly designed to be implemented in *software*. In particular, it is amenable to being implemented as a module running inside an isolated execution environment on a mobile platform. Also unlike the TPM, the MTM explicitly supports the instantiation of several parallel instances. This feature is intended to support an instance for each of a few stake-holders on a mobile platform. In principle, it could be used to support a private MTM instance to each individual software module that is loaded into an isolated execution environment.

Adding an MTM alone to a mobile platform and allowing third-party developers to access it via App-IEE APIs would serve to expose the underlying hardware security features in a uniform way across hardware platforms.

The MTM could also be used in architectures where third-party code is allowed to execute in an isolated execution environment. However, simply giving secure modules direct access to a single shared MTM instance would put all running modules into each-other's TCB; e.g., modules would be able to unseal data belonging to other modules. This limitation could be addressed by instantiating a fresh, private, MTM instance for each module that runs. Optionally, to minimize complexity and resource-usage, these on-demand MTMs could implement only a small subset of the MTM specification. This is similar to the approach taken by previous research on x86 platforms, with the MTM taking the place of the TPM [40, 46].

Another, orthogonal, way to use an MTM is for the isolated execution environment itself to use the MTM as a back-end. This strategy could provide a uniform interface for implementing the isolated execution environment itself across multiple hardware platforms.

While several researchers have implemented the MTM [20, 23, 35, 62, 65], it is not to our knowledge implemented on any off-the-shelf mobile platforms.

6.3.2 OnBoard Credentials

OnBoard Credentials (ObC) [21, 34] is an architecture to provide an isolated execution environment to third-party software modules written in the Lua[1] scripting language. It includes both App-IEE and Module-IEE APIs.

ObC provides most of the features described in Chap. 3 an isolated execution environment, secure (sealed) storage, and secure provisioning. It also provides a form

---

[1] www.lua.org

of trusted path, implemented using a management application with a customizable interface. Unfortunately it does not provide a remote attestation API, though adding one would be straightforward.

While ObC supports only Lua modules and not native-code modules, this design decision was made so that its isolated execution environment would have very small run-time memory requirements (6 KB for the Lua interpreter). This allows ObC to fit into on-chip memory, thus mitigating physical attacks such as bus-sniffing. This feature is beneficial for some use-cases, e.g., to protect the owner's secrets from being compromised if the device is physically lost or stolen, or DRM applications where the legitimate owner of the physical device may be the attacker.

ObC's key provisioning design relies heavily on the physical security of all participating devices. Secured data is provisioned or migrated between devices by encrypting it under a global program-family symmetric key. That key is, in turn, provisioned to devices trusted to protect it and to use it in accordance with ObC policy (i.e., only use it to encrypt or decrypt data for ObC programs that are part of the program-family). In this model, compromising the program-family key from any participating device is sufficient to compromise the confidentiality and integrity of data migrated by that program-family on any device—a break-once, run-anywhere attack. Hence, for applications protecting data that is confidential to the device owner, such as web site or banking credentials, it would be preferable to reduce that attack surface to the set of devices trusted by *that user*.

It may be possible to extend ObC to support a user-centric trust model, by replacing program-family-keys with user-keys, and putting the user in charge of provisioning those keys to the devices that the user owns or otherwise trusts. Such a provisioning mechanism could be built using a remote-attestation mechanism; while ObC assumes the existence of such a mechanism (using device-keys), its API does not expose a remote attestation feature to secure software modules. However, adding such an API would be straightforward.

The primary implementation of ObC uses Texas Instruments' M-Shield™ [13] to provide an isolated execution environment, secure storage, and integrity of the isolated execution environment (via secure boot). While multiple commodity smart-phones are equipped with the necessary hardware support for ObC, enabling it requires a specially signed device firmware image from the OEM or carrier, and is outside the reach of third-party developers and device owners.

6.3.3 TrustZone™ API

The TrustZone™ API (not to be confused with the TrustZone™ hardware features) is an App-IEE API for managing and invoking modules in an isolated execution environment [9].

The TrustZone™ API appears to have been designed to work with services running in the TrustZone™ secure world (Sect. 4.2.1) in particular; however, the model is fairly abstract and provides interfaces for selecting *which* secure "device" to communicate with. Hence, the TrustZone™ API could conceivably be implemented to communicate with secure services backed with other protection mechanisms, or even services running on a remote device.

The (publicly available) TrustZone™ API does *not* include Module-IEE APIs. Hence, while it could be a useful set of APIs to expose to app developers, allowing them to communicate with services running in an isolated execution environment, by itself it does not fully specify the APIs needed for *developing* such service modules.

We are not aware of any mobile platforms where the TrustZone™ API is open to third-party developers.

6.3.4 GP Trusted Execution Environment

The GlobalPlatform consortium is developing a set of standards for a Trusted Execution Environment (TEE) [55]. It includes both App-IEE APIs for applications to interact with isolated modules [53], and Module-IEE APIs for developing such modules [54].

While the system architecture specifically suggests options where the environment is created by multiplexing resources with an untrusted OS, to our knowledge the only implementations of the TEE use a dedicated device such as a Secure Element Sect. 4.3.1 or smartcard, and only run applications in the secure environment that are pre-approved by the entity deploying that device.

The TEE client specification [53] includes APIs for connecting to and invoking a secure application. The TEE internal specification [54] defines the runtime support available to secure applications running inside the TEE. These include communication with calling code outside of the TEE, secure storage (though it is unclear if state continuity is provided [43]), cryptographic primitives, and trusted time.

Of the security features from Chap. 3, those missing are remote attestation, secure provisioning, and trusted path. In principle remote attestation can be added, which, as discussed in Sect. 3.3, can be used to build secure provisioning.

Chapter 7
Analysis and Recommendations

We now give our analysis of the security properties that today's mobile devices can provide, and offer recommendations to the research community, to app developers, to platform integrators, and to hardware vendors.

The set of primary stake-holders today includes only the OEMs and telecommunications carriers (and their immediate business partners). Thus, the hardware security primitives that are actually included in mass-market mobile devices are only those of interest to the OEMs and telecommunications providers. It is our primary recommendation that application developers and device owners be considered first-class *stake-holders* by OEMs and telecommunications service providers. While economics may prevent the inclusion of additional hardware security primitives in mass-market devices without a compelling business reason, those primitives which are present should be leveraged to offer additional security features to application developers and devices owners.

We have shown that—while helpful—the security APIs provided to application developers by today's mobile OSes are inadequate because of the continuing ease with which mobile device OSes are compromised. We have also shown (Chap. 2) that the market has responded to the need for security features with add-on hardware that provides Secure Element functionality, either exclusively or in conjunction with new I/O interfaces. Some newer devices (e.g., the Nexus S) are beginning to include embedded Secure Elements; unfortunately, this hardware is being monopolized by a single application.

Given the rise in add-on security devices, it is reasonable to question why the OEMs and carriers have not responded more aggressively by opening up or including additional security features. On this issue we can only speculate, but we list here a few plausible explanations: (1) an existing culture of security-through-obscurity is reluctant to embrace change; (2) business interests are attempting to corner the market for their exclusive use; (3) fragmentation in existing proposals hampers their adoption for fear of future incompatibilities; or (4) there is little awareness of the level of maturity for proposals for multiplexing hardware security features between multiple stake-holders.

A. Vasudevan et al., *Trustworthy Execution on Mobile Devices*, SpringerBriefs in Computer Science, DOI: 10.1007/978-1-4614-8190-4_7, © The Author(s) 2014

We reject the position that OEMs or carriers are unable to monetize such additions. Features sell phones, and security primitives enable application developers to produce new kinds of applications. We reject the position that DRM is the only viable use for such features. Digital media is inherently break-once, run-anywhere, and it *will* eventually be broken. We argue that hardware-backed security features can immediately add significant value in such areas as protecting users' cached login credentials and encrypting users' data while at rest on the phone. Mobile payments and myriad other applications can follow in quick succession.

7.1 Research Community Recommendations

It is our recommendation to the research community to continue to investigate viable architectures for multiplexing mutually-distrusting stake-holders on resource-constrained hardware security primitives (Chap. 6). This is especially important as virtualization extensions make their way to the ARM architecture (Sect. 4.2.2), opening up the possibility for two divergent approaches (split-world versus virtualization). Special attention should be paid to the possibility for a heterogeneous threat model: OEMs and carriers are concerned about defenses against physical attacks, whereas many use-cases for protecting the end-user's data are primarily concerned with software-based attacks that arrive via a network connection.

Development hardware with a multitude of unlocked security features is now readily available and inexpensive (Sect. 4.8). Though hardware with virtualization extensions remains unavailable at the time of this writing, ARM's toolkit enables emulation of Cortex A15 platforms today. Open-source contributions of viable multiplexing architectures can serve to raise awareness with OEMs, carriers, and application developers. We are optimistic that a robust reference implementation could even enjoy widespread deployment.

The fear of fragmentation of security APIs can be addressed by developing consistent interfaces. We recommend the adoption of consistent Module-IEE and App-IEE APIs, so that application developers that endeavor to privilege-separate their programs today can continue to reap the security benefits into the future without significant risk of incompatibility or maintenance / support nightmares. Support for multiple feature sets may also be reasonable. For example, credential programs such as those enabled by OnBoard Credentials (Sect. 6.3.2) may reasonably coexist with more feature-rich isolated execution environments that allow arbitrary computation (within resource limits).

7.2 Application Developer Recommendations

It is our recommendation to application developers to continue to demand improved security APIs and primitives in the development environment for popular mobile device platforms. The incredible volume of misinformation bandied about on Internet forums about how to protect cached credentials, encrypt data at rest, or give users a false sense of security is deeply disturbing.

We encourage application developers to learn about existing proposals for Module-IEE and App-IEE APIs, and to consider their implications for the architecture of their applications. Especially those developers with an interest in open-source can produce reference implementations that we expect may be rapidly adopted by other developers.

7.3 Platform Integrator Recommendations

We recommend that platform integrators (typically network carriers) take an interest in the security of applications on their devices. We argue that they should adopt a realistic perspective regarding the robustness of the OS APIs for security.

Hardware-backed or otherwise isolated security features are in demand by application developers. Existing Module-IEE and App-IEE proposals should be adopted, to avoid fragmentation and a lack of developer buy-in. These security features will enable application developers to add new value to the mobile device platforms as a whole, resulting in an overall increase in the utility of mobile devices.

We strongly urge platform integrators to make hardware security features available that are otherwise included in the silicon but disabled immediately during every boot. Other developers and the open-source community are likely to energize and do much of the engineering. As a viable first step, we recommend an implementation of the TCG's Mobile Trusted Module (MTM) in devices with TrustZone™ capabilities that are otherwise unused (Sect. 6.3.1). This suggestion is consistent with the App-IEE-only approach discussed in (Sect. 6.1), and offers new security features to application developers. Note that it does not give application developers the ability to directly execute their own code inside of an isolated execution environment (Sect. 3.1) and (Sect. 6.2). Thus, it is a reasonable compromise between conservative, risk-averse OEMs and carriers, and a useful set of APIs for application developers.

We emphasize that platform openness and security are *not* fundamentally opposing trade-offs, and that additional access to hardware security primitives will not somehow weaken them. The key to reconciling this common misconception is isolation. A single vulnerable application—even a security-critical subcomponent of an application—should never be in a position where it is able to compromise the entire device.

7.4 Hardware Vendor Recommendations

Unconstrained memory isolation and improved protection against DMA-based attacks (Sect. 4.7) are significant needs in current device hardware. It is more difficult for us to justify the added expense in device hardware at the present time. If the market does indeed parallel our recommendations in the preceding sections, and existing hardware security features begin to enable new applications, then the logical next step is to offer additional hardware security features.

To this end, our recommendation is to address the DMA insecurity problem (Sect. 4.7). This will not only add protection against currently prevalent attacks from malicious peripherals [61], but will also result in the automatic inclusion of memory address-space controllers such as a TZASC and/or TZMA (Sect. 4.2.1), so that security-sensitive modules that execute in isolation need not grapple with today's dearth of Tightly Coupled Memory.

Chapter 8
Summary

We are now in the post-PC era, yet our mobile devices are insecure. In this book, we consider the different stake-holders in today's mobile device ecosystem, and analyze why widely-deployed hardware security primitives on mobile device platforms are inaccessible to application developers and end-users. We systematize existing proposals for leveraging such primitives, and show that they can indeed strengthen the security properties available to applications and users, all without reducing the properties currently enjoyed by OEMs and network carriers. We also highlight shortcomings of existing proposals and make recommendations for future research that may yield practical, deployable results.

A. Vasudevan et al., *Trustworthy Execution on Mobile Devices*, SpringerBriefs
in Computer Science, DOI: 10.1007/978-1-4614-8190-4_8, © The Author(s) 2014

References

1. Android: an open handset alliance project—issue 10809: password is stored on disk in plain text. http://code.google.com/ (2010)
2. Android Developers: Android API: AccountManager. http://developer.android.com/. Accessed (2011)
3. Android Open Source: Notes on the implementation of encryption in Android 3.0. http://source.android.com. Accessed (2011)
4. Android.com: Android 4.0 platform highlights. http://developer.android.com/sdk/android-4.0-highlights.html
5. Andrus, J., Dall, C., Hof, A.V., Laadan, O., Nieh, J.: Cells: a virtual mobile smartphone architecture. In: Proceedings of the 23rd ACM Symposium on Operating Systems Principles (SOSP) (2011)
6. Apple: iOS: understanding data protection. Article HT4175 (2011)
7. ARM Limited: ARM builds security foundation for future wireless and consumer devices. ARM Press Release (2003)
8. ARM Limited: ARM security technology: building a secure system using TrustZone technology. WhitePaper PRD29-GENC-009492C (2009)
9. ARM Limited: TrustZone API specification 3.0. Technical report PRD29-USGC-000089 3.1, ARM (2009)
10. ARM Limited: AMBA 4 AXI4-stream protocol version 1.0 specification (2010)
11. ARM Limited: AMBA APB protocol version 2.0 specification (2010)
12. ARM Limited: Virtualization extensions architecture specification. http://infocenter.arm.com (2010)
13. Azema, J., Fayad, G.: M-Shield mobile security: making wireless secure. Texas Instruments WhitePaper (2008)
14. Becher, M., Freiling, F.C., Hoffman, J., Holz, T., Uellenbeck, S., Wolf, C.: Mobile security catching up? revealing the nuts and bolts of the security of mobile devices. In: Proceedings of the IEEE Symposium on Security and Privacy (2011)
15. Choudhary, B., Risikko, J.: Mobile device security element: Key findings from technical analysis v1.0. In: Mobey Forum: Mobile Financial Services. http://www.mobeyforum.org/Press-Documents/White-papers (2005)
16. Comex: JailbreakMe. http://jailbreakme.com/. Accessed (2011)
17. Costan, V., Sarmenta, L.F., van Dijk, M., Devadas, S.: The trusted execution module: commodity general-purpose trusted computing. In: Proceedings of the CARDIS (2008)
18. Cranor, L.F.: What do they "indicate"?: evaluating security and privacy indicators. Interactions 13(3), 45–47 (2006)
19. Dall, C., Nieh, J.: Kvm for arm. In: Proceedings of the 12th Annual Linux, Symposium (2010)

20. Dietrich, K., Winter, J.: Towards customizable, application specific mobile trusted modules. In: Proceedings of the ACM workshop on Scalable Trusted Computing (2010)
21. Ekberg, J.E., Asokan, N., Kostiainen, K., Rantala, A.: Scheduling execution of credentials in constrained secure environments. In: Proceedings of the ACM Workshop on Scalable Trusted Computing (2008)
22. Ekberg, J.E., Kylänpää, M.: Mobile trusted module (mtm): an introduction. Technical report NRC-TR-2007-015, Nokia Research Center (2007)
23. Ekberg, J.E., Kylänpää, M.: MTM implementation on the TPM emulator. Source code. http://mtm.nrsec.com (2008)
24. ElcomSoft: Proactive software: iOS forensic toolkit (2011)
25. Gilbert, P., Cox, L.P., Jung, J., Wetherall, D.: Toward trustworthy mobile sensing. In: Proceedings of the Workshop on Mobile Computing Systems and Applications (HotMobile) (2010)
26. Gligor, V.D., Chandersekaran, C.S., Chapman, R.S., Dotterer, L.J., Hecht, M.S., Jiang, W.D., Johri, A., Luckenbaugh, G.L., Vasudevan, N.: Design and implementation of secure Xenix. IEEE Trans. Software Eng. **13**, 208–221 (1986)
27. GottaBeMobile: Texas instruments ARM OMAP4 becomes first mobile CPU to get Netflix certification for android HD streaming. http://gottabemobile.com/ (2011)
28. Green Hills Software: Emergence of the mobile multivisor. http://ghs.com/ (2011)
29. Guaus, J., Kanniainen, L., Koistinen, P., Laaksonen, P., Murphy, K., Remes, J., Taylor, N., Welin, O.: Best practice for mobile financial services v1.0. Technical report, Mobey Forum. http://www.mobeyforum.org/Press-Documents/White-papers (2008)
30. Hecht, M.S., Carson, M.E., Chandersekaran, C.S., Chapman, R.S., Dotterrer, L.J., Gligor, V.D., Jiang, W.D., Johri, A., Luckenbaugh, G.L., Vasudevan, N.: UNIX without the superuser. In: Proceedings of the USENIX Technical Conference, pp. 243–256 (1987)
31. Heider, J., Boll, M.: Lost iPhone? lost passwords! practical consideration of iOS device encryption security. Technical report, Fraunhofer SIT (2011)
32. Klein, G., Elphinstone, K., Heiser, G., Andronick, J., Cock, D., Derrin, P., Elkaduwe, D., Engelhardt, K., Kolanski, R., Norrish, M., Sewell, T., Tuch, H., Winwood, S.: seL4: formal verification of an OS kernel. In: Proceedings of the ACM Symposium on Operating Systems Principles (SOSP) (2009)
33. Koistiainen, K., Reshetova, E., Ekberg, J.E., Asokan, N.: Old, new, borrowed, blue-a perspective on the evolution of mobile platform security architectures. In: Proceedings of the 1st ACM Conference on Data and Application Security and Privacy (CODASPY) (2011)
34. Kostiainen, K., Ekberg, J.E., Asokan, N., Rantala, A.: On-board credentials with open provisioning. In: Proceedings of the ASIACCS (2009)
35. Kursawe, K., Schellekens, D.: Flexible MicroTPMs through disembedding. In: Proceedings of the ASIACCS (2009)
36. Lampson, B.: Usable security: how to get it. Commun. ACM **52**(11) (2009)
37. Lineberry, A., Strazzere, T., Wyatt, T.: Inside the android security patch lifecycle. Presented at BlackHat (2011)
38. Mastin, M.: Square versus intuit gopayment: mobile credit card systems compared. In: PCWorld. http://www.pcworld.com/businesscenter/article/239250/ (2011)
39. McCammon, R.: How to build a more secure smartphone with mobile virtualization and other commercial off-the-shelf technology. Open Kernel Labs Technology White Paper (2010)
40. McCune, J.M., Li, Y., Qu, N., Zhou, Z., Datta, A., Gligor, V., Perrig, A.: TrustVisor: efficient TCB reduction and attestation. In: Proceedings of the IEEE Symposium on Security and Privacy (2010)
41. Mills, E.: Researchers find avenues for fraud in square. In: CNET. http://news.cnet.com/8301-27080_3-20088441-245/ (2011)
42. Open Kernel Labs: OK labs company datasheet. http://www.ok-labs.com (2010)
43. Parno, B., Lorch, J.R., Douceur, J.R., Mickens, J., McCune, J.M.: Memoir: Practical state continuity for protected modules. In: Proceedings of the IEEE Symposium on Security and Privacy (2011)

44. Popek, G.J., Goldberg, R.P.: Formal requirements for virtualizable third generation architectures. Commun. ACM **17** (1974)
45. Reveilhac, M., Pasquet, M.: Promising secure element alternatives for NFC technology. In: Proceedings of the International Workshop on Near Field Communication (NFC) (2009)
46. Sailer, R., Jaeger, T., Valdez, E., Cáceres, R., Perez, R., Berger, S., Griffin, J., van Doorn, L.: Building a MAC-based security architecture for the Xen opensource hypervisor. In: Proceedings of the Annual Computer Security Applications Conference (2005)
47. Saroiu, S., Wolman, A.: I am a sensor, and I approve this message. In: Proceedings of the Workshop on Mobile Computing Systems and Applications (HotMobile) (2010)
48. Schechter, S.E., Dhamija, R., Ozment, A., Fischer, I.: The emperor's new security indicators. In: IEEE Symposium on Security and Privacy (2007)
49. Schell, S.V., Narang, M., Caballero, R.: US patent 2011/0269423 Al: wireless network authentication apparatus and methods (2011)
50. Schwartz, M.J.: Apple iOS zero-day PDF vulnerability exposed. In: InformationWeek. http://www.informationweek.com/news/231001147 (2011)
51. Sun Microsystems, Inc.: Java card specifications v3.0.1: classic edition, connected edition (2009)
52. TCG Mobile Phone Working Group: TCG mobile trusted module specification, version 1.0, revision 7.02 (2010)
53. Global Platform Device Technology: TEE client API specification version 1.0. Technical report GPD_SPE_007. http://globalplatform.org/ (2010)
54. GlobalPlatform Device Technology: TEE internal API specification version 0.27. Technical report GPD_SPE_010. http://globalplatform.org/ (2011)
55. GlobalPlatform Device Technology: TEE system architecture version 0.4. Technical report GPD_SPE_009. http://globalplatform.org/ (2011)
56. Texas Instruments E2E Community: Setup of secure world environment using TrustZone: OMAP35X processors forum. http://e2e.ti.com/ (2010)
57. US Department of Defense: Trusted computer system evaluation criteria (orange book). DoD 5200.28-STD (1985)
58. Vasudevan, A., Chaki, S., Jia, L., McCune, J., Newsome, J., Datta, A.: Design, implementation and verification of an extensible and modular hypervisor framework. In: Proceedings of the IEEE Symposium on Security and Privacy (S&P) (2013)
59. Vasudevan, A., Owusu, E., Zhou, Z., Newsome, J., McCune, J.M.: Trustworthy execution on mobile devices: what security properties can my mobile platform give me? In: Proceedings of the 5th International Conference on Trust and Trustworthy Computing (Trust 2012) (2012)
60. Virtual Open Systems: Multi-core smp kvm full virtualization on cortex-a15 platforms. http://www.virtualopensystems.com (2013)
61. Wang, Z., Stavrou, A.: Exploiting smart-phone usb connectivity for fun and profit. In: Proceedings of the Annual Computer Security and Applications Conference (ACSAC) (2010)
62. Winter, J.: Trusted computing building blocks for embedded linux-based ARM TrustZone platforms. In: Proceedings of the ACM Workshop on Scalable Trusted Computing (2008)
63. Xen.org: Xen ARM project. http://wiki.xen.org/wiki/XenARM. Accessed (2011)
64. Yao, Y.: Security issue exposed by android accountmanager. http://security-n-tech.blogspot.com/2011/01/security-issue-exposed-by-android.html (2011)
65. Zhang, X., Aciicmez, O., Seifert, J.P.: A trusted mobile phone reference architecture via secure kernel. In: Proceedings of the ACM Workshop on Scalable Trusted Computing (2007)
66. Zhou, Z., Gligor, V.D., Newsome, J., McCune, J.M.: Building verifiable trusted path on commodity x86 computers. In: Proceedings of the IEEE Symposium on Security and Privacy (2012)

About the Author

Amit Vasudevan (Bio)

Amit Vasudevan is a Research Scientist at CyLab, Carnegie Mellon University. He received his Ph.D. and M.S degrees from the Computer Science Department at UT Arlington and spent three years as a Post-doc at Carnegie Mellon University. Before that, he obtained his B.E. from the Computer Science Department at the BMS College of Engineering, India. His research interests include secure systems, virtualization, trusted computing, malware analysis and operating systems. His present research focuses on hypervisor-based trustworthy code execution and formal verification methodologies. He is the principal force behind the design and development of the eXtensible and Modular Hypervisor Framework (http://xmhf.org)—an open-source, clean, bare-bones, formally verifiable hypervisor framework—which forms the foundation for a new class of (security-oriented) hypervisor-based applications ("hypapps").

Jonathan M. McCune (Bio)

Jonathan McCune is a Research Systems Scientist for CyLab at Carnegie Mellon University. He earned his Ph.D. degree in Electrical and Computer Engineering from Carnegie Mellon University, and received the A.G. Jordan thesis award. He received his B.Sc. degree in Computer Engineering from the University of Virginia (UVA). Jonathan's research interests include secure systems, trusted computing, virtualization, and spontaneous interaction between mobile devices. When keyboards and LCDs get to be too much for him, Jon can usually be found riding a bike.

James Newsome (Bio)

James Newsome joined CyLab as a Research Scientist in 2010. James received his B.Sc. from the University of Michigan in 2002 and his Ph.D. from Carnegie Mellon University in 2008. He was a Research Engineer at Bosch Research Technology Center from 2008–2010, where he analyzed and designed security mechanisms in embedded and distributed systems. James's previous and current research includes automated software-exploit detection, analysis, signature-generation, and patch-generation; binary analysis; information flow; secure distributed and embedded system design; trusted computing; and isolated code execution.

Curriculum Vitae

Amit Vasudevan

Contact Information

Research Systems Scientist
CyLab, Carnegie Mellon University
4819 27th Road South, Arlington, VA 22206, USA
Phone: +1-571-329-6340
E-mail: amitvasudevan@acm.org
WWW: http://hypcode.org

Research Interests

- Trusted Computing and Virtualization Security
  - Secure hypervisors and hypervisor verification methodologies
  - Trustworthy execution and execution provenance/logging
- Security of mobile devices including mobile phones, laptops and tablets
- Malware (malicious-code) analysis and detection
- Operating system security and kernel architecture

Teaching Interests

Computer Architecture and Assembly Language Programming, C Programming, Systems Programming, Virtualization/Hypervisors, Trusted Computing, Malware Analysis, Operating Systems, Embedded Systems.

A. Vasudevan et al., *Trustworthy Execution on Mobile Devices*, SpringerBriefs
in Computer Science, DOI: 10.1007/978-1-4614-8190-4, © The Author(s) 2014

Education

The University of Texas, Arlington, Arlington, TX, USA

Ph.D., Computer Science and Engineering, May 2007

- Thesis title: *WiLDCAT: An Integrated Stealth Environment for Dynamic Malware Analysis*
- Advisor: Prof. Ramesh Yerraballi
- GPA: 4.0/4.0

M.S., Computer Science and Engineering, December 2003

- Thesis title: *Sakthi: A Retargetable Dynamic Framework for Binary-Instrumentation*
- Advisor: Prof. Ramesh Yerraballi
- GPA: 4.0/4.0

BMS College of Engineering (Bangalore University), Bangalore, India

B.E., Computer Science and Enginering, September 2001

- GPA: First Class with Distinction, 3.9/4.0

Academic Appointments

Carnegie Mellon University
Research Systems Scientist   **October 2010 to present**
CyLab

Responsibilities include basic research in the field of computer security with an emphasis on trustworthy computing, virtualization security and embedded/mobile virtualization; software development, and solicitation of research funding. Active research projects:

- eXtensible, Modular Hypervisor Framework (XMHF)
- Hypervisor Verification
- Hypervisor-based Verifiable Platform Resource Accounting
- Isolated Execution on Mobile Devices
- On-CPU Isolation and Root-of-Trust

Active open-source software development:

- xmhf.org. An extensible, modular and formally verifiable hypervisor framework for x86 systems (AMD and Intel) with support for dynamic root of trust, hypervisor boot integrity measurement and isolated execution for security-sensitive code.

Carnegie Mellon University
Postdoctoral Researcher   **September 2007 to October 2010**
CyLab

Responsibilities include research in virtualization security, malware analysis and trustworthy computing; software development, and solicitation of research funding.
Research projects include:

- Lockdown: Towards a Safe and Practical Architecture for Security Applications
- Secure Execution Trace Recording
- SecVisor: Kernel-mode Execution Integrity
- Tracking Unknown (0-day) Malware

The University of Texas at Arlington
Assistant Instructor   **January 2004 to May 2007**
Department of Computer Science and Engineering

Taught undergraduates CSE2312: Computer Organization and Assembly Language Programming and CSE1301: Introduction to C Programming. Set course syllabus, authored course lectures, designed exams and assignments.

Professional Experience

NoFuss Security, Inc, Pittsburgh, PA, USA
Senior Security Architect   **February 2011 to present**

Member of three-person engineering team: design, development, and support for commercialization of trustworthy computing technologies. Principal investigator on a project on analysis of COTS hypervisor security. Major tasks included systematic attack surface enumeration, discovery of design-level vulnerabilities and constructing mitigation approaches.

Cognizant Technologies and Solutions Corp., India
Programmer Analyst   **January 2002 to May 2002**

Developed a bridge framework (native C to Enterprise Java Beans) for faster porting of existing native Solaris/C application containers to Enterprise Java Beans.

Bharat Electronics, India
Research Intern   **June 2001 to December 2001**

Designed software control mechanisms and a command and control center for monitoring and controlling a group of remote naval radar units.

Books and Book Chapters

1. Amit Vasudevan, Jonathan M. McCune, James Newsome. "Trustworthy execution on mobile devices". What security properties can my mobile platform give *me*?. In preparation. Springer Briefs. **(Invited)**
2. Amit Vasudevan. "Effective Malware Analysis Using Stealth Breakpoints". In Strategic and Practical Approaches for Information Security Governance: Technologies and Applied Solutions. IGI Global, ISBN-10: 1466601973, 2012 **(Invited)**

Refereed Publications

1. Amit Vasudevan, Sagar Chaki, Limin Jia, Jonathan McCune, Jim Newsome, Anupam Datta. "Design, Implementation and Verification of an eXtensible and Modular Hypervisor Framework". In IEEE Symposium on Security and Privacy, 2013. To appear.
2. Chen Chen, Petros Maniatis, Adrian Perrig, Amit Vasudevan, Vyas Sekar. "Towards Verifiable Resource Accounting for Outsourced Computation". In ACM Virtual Execution Environments, 2013. To appear.
3. Carsten Willems, Ralf Hund, Amit Vasudevan, Andreas Fobian, Dennis Felsch, Thorsten Holz. "Down to the Bare Metal: Using Processor Features for Binary Analysis". In IEEE Annual Computer Security and Applications Conference (ACSAC), 2012.
4. Amit Vasudevan, Bryan Parno, Ning Qu, Virgil D. Gligor, Adrian Perrig. "Lockdown: Towards a Safe and Practical Architecture for Security Applications on Commodity Platforms". In International Conference on Trust and Trustworthy Computing (TRUST), Vienna, Austria, 2012.
5. Amit Vasudevan, Emmanuel Owusu, Zongwei Zhou, James Newsome, Jonathan M. McCune. "Trustworthy Execution on Mobile Devices: What Security Properties Can My Mobile Platform Give Me?". In International Conference on Trust and Trustworthy Computing (TRUST), Vienna, Austria, 2012.
6. Amit Vasudevan, Jonathan McCune, James Newsome, Adrian Perrig, Leendert van Doorn. "CARMA: A Hardware Tamper-Resistant Isolated Execution Environment on Commodity x86 Platforms". In ACM Symposium on Information, Computer and Communications Security (ASIACCS), 2012.
7. Jason Franklin, Sagar Chaki, Anupam Datta, Jonathan M. McCune, Amit Vasudevan. "Parametric Verification of Address Space Separation". In First Conference on Principles of Security and Trust (POST) 2012. **(Best Paper Nomination)**
8. Amit Vasudevan, Ning Qu and Adrian Perrig, "XTRec: Secure Real-time Execution Trace Recording on Commodity Platforms". In 44th Hawaii International Conference in System Sciences (HICSS), Hawaii, 2011. **(Best Paper Nomination)**

9. Amit Vasudevan, Jonathan M. McCune, Ning Qu, Leendert van Doorn and Adrian Perrig, "Requirements for an Integrity-Protected Hypervisor on the x86 Hardware Virtualized Architecture". In 3rd International Conference on Trust and Trustworthy Computing (TRUST), Berlin, Germany, 2010.
10. Amit Vasudevan, "Reinforced Stealth Breakpoints". In 4th IEEE Conference on Risks in Internet Systems (CRiSIS), Toulouse, France, 2009.
11. Amit Vasudevan, "MalTRAK: Tracking and Eliminating Unknown Malware". In IEEE 24st Annual Computer Security and Applications Conference (ACSAC), Anaheim, CA, 2008.
12. Amit Vasudevan and Ramesh Yerraballi, "Cobra: Fine-grained Malware Analysis using Stealth Localized-executions". In 2006 IEEE Symposium on Security and Privacy, Oakland, CA.
13. Amit Vasudevan and Ramesh Yerraballi, "SPiKE: Engineering Malware Analysis Tools using Unobtrusive Binary-Instrumentation". In 29th Australasian Conference in Computer Science (ACSC), Hobart, Australia, 2006. (**Best Paper Nomination**)
14. Amit Vasudevan and Ramesh Yerraballi, "Stealth Breakpoints". In IEEE 21st Annual Computer Security and Applications Conference (ACSAC), Tucson, AZ, 2005.
15. Ashish Chawla, Ramesh Yerraballi and Amit Vasudevan, "Coalesced QoS: A Pragmatic Approach to a Unified Model to Support Quality Of Service (QoS) in High Performance Kernel-Less Operating System (KLOS)". In Advances in Systems, Computing Sciences and Software Engineering: Proceedings of SCSS 2005 (14), December 2005. ISBN-10: 1-4020-5262-6.
16. Amit Vasudevan, Ramesh Yerraballi and Ashish Chawla, "A High Performance Kernel-Less Operating System Architecture". In 28th Australasian Conference in Computer Science (ACSC), New Castle, Australia, 2005.
17. Amit Vasudevan and Ramesh Yerraballi, "SAKTHI A Retargetable Dynamic Framework for Binary Instrumentation". In 2004 Hawaii International Conference on Computer Sciences (HICCS), Honolulu, HI, 2004. ISSN:1545-672.
18. Amit Vasudevan, Ramesh Yerraballi and Ashish Chawla, "KLOS A High Performance Kernel-Less Operating System". In 2003 IEEE RTSS Work in Progress.

Technical Reports

1. Sagar Chaki, Amit Vasudevan, Limin Jia, Jonathan M. McCune, and Anupam Datta. "Design, Development and Automated Verification of an Integrity-Protected Hypervisor". CMU CyLab Technical Report CMU-CyLab-12-017. July 2012.
2. Amit Vasudevan, Jonathan M. McCune, and James Newsome. "Design and Implementation of an eXtensible and Modular Hypervisor Framework". CMU CyLab Technical Report CMU-CyLab-12-014. June 2012.

3. Jason Franklin, Sagar Chaki, Anupam Datta, Jonathan Mccune and Amit Vasudevan. "Parametric Verification of Address Space Separation". CMU Cylab Technical Report CMU-CyLab-12-001. January 2012.
4. Amit Vasudevan, Emmanuel Owusu, Zongwei Zhou, James Newsome, and Jonathan McCune. "Trustworthy Execution on Mobile Devices: What security properties can my mobile platform give me?". CMU CyLab Technical Report CMU-CyLab- 11-023. November 2011.
5. Amit Vasudevan, Bryan Parno, Ning Qu and Adrian Perrig. "Lockdown: A Safe and Practical Environment for Security Applications". CMU CyLab Technical Report CMU-CyLab-09-011. July 2009.
6. Amit Vasudevan, Ning Qu, Adrian Perrig. "XTREC: Secure Realtime Instruction-level Control Flow Recording on Commodity Platforms". CMU CyLab Technical Report CMU-CyLab-09-007. March 2009.

Patents

1. Application Number 61/273,454. Inventor(s): Virgil D. Gligor, Adrian Perrig, Anupam Datta, Jonathan M. McCune, Ning Qu, Bryan J. Parno, Amit Vasudevan, Yanlin Li (equal rights). "User-Verifiable Execution of Security-Sensitive Code on Untrusted Platforms", Filing Date: 04/2009 (Pending).
2. Application Number 60/861,621. Inventor(s): Amit Vasudevan. "Method and System for Stealth Runtime Fine and Coarse-grained Malware Analysis", Filing Date: 11/28/2006 (Pending).

Grants

1. "COTS Hypervisor Security: Architectural Analysis", United States Air Force, 2011. [Role: PI, $50,000]
2. "Real-time Execution Trace Recording and Analysis on Commodity Platforms", Northrup Grumman, 2009–2010. [Role: Co-PI, $200,000]
3. "Hypervisor-based Secure Virtualization", United States Air Force, 2008–2009. [Role: Lead Researcher, $170,000]

Teaching Experience

The University of Texas at Arlington, Arlington, TX

Assistant Instructor **January 2004 to May 2007**

- Instructor for CSE 2312: Computer Organization and Assembly Language Programming
  - Responsible for course lecture (3 hours/week), designing and grading exams and assignments
  - Students learnt low-level details of x86 platform architecture, devices and assembly language programming in the context of various hands-on programming assignments
- Instructor for CSE 1301: Introduction to C Programming
  - Responsible for course lecture (3 hours/week), designing and grading exams and assignments
  - Students learnt various functional aspects of the C programming language including data types, input and output functions, structures, pointers, arrays, file handling, calling conventions with hands-on programming assignments

Teaching Assistant **May 2006 to August 2006**

- CSE5306: Operating Systems II
  - Tutored students one-to-one and proctored exams
  - Graded assignments and class project deliverables

Teaching Assistant **August 2002 to January 2004**

- CSE2312: Computer Organization and Assembly Language Programming
  - Tutored students one-to-one, proctored and graded exams
  - Prepared and graded assignments and class project deliverables

Hardware and Software Skills

Programming Languages/Tools:

- Assembly (x86 and ARM, 32/64-bit), C, C++, Java, JavaScript, Pascal, Perl, UNIX shell scripting, GNU make, and others

Hypervisor Development:

- x86 and ARM hypervisor design and implementation: AMD Secure Virtual Machine (SVM), Intel Virtualization Technology (VT), ARM TrustZone (TZ) and Virtualization Extension Architecture Specfication (VEAS)

- Dynamic Root-of-Trust/High-Assurance Boot, Nested and Shadow Paging, I/O Virtualization, IOMMU (VT-d,TZPC) and SMM/TZ mode containers
- Principal force behind the design, implementation and verification of the eXtensible and Modular Hypervisor Framework (http://xmhf.org). Also explored internals of other open-source hypervisors such as KVM and Xen

System-level Development:

- x86 and embedded ARM architectures: Trusted Platform Module (TPM), custom BIOS development (e.g., coreboot), protected-mode, debugging/SMM/TZ, hard- ware debug tools, PCI/PCI-E, power management (APM/ACPI), multi-core (SMP), device interrupt interfaces (APIC/IOAPIC/GIC) and others.
- Device drivers: Linux Loadable Kernel Modules (LKM), Windows kernel-mode legacy/WDM drivers, NDIS virtual miniports, and file-system filter drivers
- Linux/Windows kernel development, kernel instrumentation, Windows native applications, disassemblers and emulators.

Software Verification Tools:

- CBMC, SATABS, Wolverine and BLAST

Version Control and Software Configuration Management:

- DVCS (Mercurial, Git) and VCS (SVN)

  Embedded and Real-time Systems:

- Software and hardware development with several MCU and DSP platforms (e.g., Freescale i.MX53, Atmel ATmega MCU's, LPC2148 MCU's, and others)

Operating Systems:

- Microsoft Windows family, Android, Linux

Productivity Applications:

- TEX (LATEX, BibTEX, PSTricks), Vim, most common productivity packages (for Windows and Linux platforms)

Honors and Awards

- Whos Who Among Students in American Universities and Colleges, University of Texas at Arlington (2005).
- University of Texas at Arlington University Scholar Award, University of Texas at Arlington. (2004 and 2005)
- Cyneta Networks Outstanding Graduate Teaching Assistant Award, Dept. of CSE, University of Texas at Arlington. (2004)
- Graduate Teaching Assistantship, Dept. of CSE, University of Texas at Arlington. (2002–2004)

- Masters Dean Fellowship, University of Texas at Arlington. (2002–2004)
- National Science Talent Search Examination Prize. India. (1996)
- POWERS, Merit Scholarship, SSLC board, India. (1994)

Talks

- Trustworthy Execution on Mobile Devices: What Security Properties Can My Mobile Platform Give Me?. TRUST, June 2012
- Lockdown: Towards a Safe and Practical Architecture for Security Applications on Commodity Platforms. TRUST, June 2012
- Requirements for an Integrity-Protected Hypervisor on the x86 Hardware Virtualized Architecture. TRUST, June 2010
- Enterprise Security Considerations/ Secure Practices for Distributed Computing and Virtualization. Lockheed Martin Technical Leadership Meet, New Orleans, USA, March 2009. **(Invited)**
- MalTRAK: Tracking and Eliminating Unknown Malware. IEEE ACSAC, December 2008
- WiLDCAT: An Integrated Stealth Environment for Dynamic Malware Analysis. Cy- Lab Seminar Series, CyLab, Carnegie Mellon University, USA, July 2007. **(Invited)**
- Cobra: Fine-grained Malware Analysis using Stealth Localized-executions. IEEE SP, May 2006.
- SPiKE: Engineering Malware Analysis tools using Unobtrusive Binary Instrumentation. ACSC, January 2006
- Stealth Breakpoints. IEEE ACSAC, December 2005.
- SAKTHI: A Retargetable Dynamic Framework for Binary Instrumentation. HICCS, January 2004.

Professional Service

PC Member: International Conference on Availability, Reliability and Security (AReS) 2013

PC Member: IEEE International Conference on Risks and Security of Internet and Systems (CRiSIS) 2012

PC Member: IEEE International Conference on Risks and Security of Internet and Systems (CRiSIS) 2011

PC Member: Conference on Decision and Game Theory for Security (GameSec) 2010

PC Member: IEEE International Conference on Risks and Security of Internet and Systems (CRiSIS) 2010

PC Member: Asia-Pacific Signal and Information and Signal Processing Association—Annual Conference 2009

Referee Service

- *ACM Transactions on Computer Systems (TOCS)*
- *Journal of Systems and Software (JSS)*
- *ACM Symposium on Operating System Principles (SOSP)*
- *ACM Conference on Computers and Communication Systems (CCS)*
- *European Symposium on Research In Computer Security (ESORICS)*
- *Architectural Support for Programming Languages and Operating Systems (ASPLOS)*
- *Network and Distributed Systems Symposium (NDSS)*
- *Operating System Design and Implementation (OSDI)*

Professional Memberships

Institute for Electrical and Electronics Engineers (IEEE), Member, 2007–present

Assocation of Computing Machinery (ACM), Member, 2007–present

Jonathan M. McCune

Personal Details

Name: Jonathan M. McCune
Address: ECE and Cylab,
Carnegie Mellon University,
4720 Forbes Ave.,
2126 CIC Pittsburgh,
PA 15213 USA
Phone: +1-412-2683992
E-mail: jonmccune@cmu.edu,
Website: http://www.ece.cmu.edu/~jmmccune

Specialties

Operating system security, virtualization, trustworthy computing (e.g., TCG), security of mobile and wireless devices, including mobile phones, laptops, tablets, and sensor networks.

Education

Ph.D. in Electrical and Computer Engineering (January 2009)
Carnegie Mellon University, Pittsburgh, PA USA
Title: *Reducing the Trusted Computing Base for Applications on Commodity Systems.*
Advisors: Adrian Perrig and Michael K. Reiter
Recipient of the A. G. Jordan Award for outstanding research and service within CMU ECE

M.S. in Electrical and Computer Engineering (May 2005)
Carnegie Mellon University, Pittsburgh, PA USA

B.S. in Computer Engineering with High Distinction (May 2003)
University of Virginia, Charlottesville, VA USA

Academic Experience

Research Systems Scientist (February 2009–Present)
Carnegie Mellon University, Pittsburgh, PA USA
Responsibilities include basic research, software development, and solicitation of research funding.
Active research projects:

- eXtensible, Modular Hypervisor Framework (XMHF)
- TrustVisor: Effcient TCB Reduction and Attestation
- Flicker: Minimal TCB Code Execution
- Isolated Execution on Mobile Devices
- Embedded Processor Root of Trust
- Datacenter Applications of Integrity Measurement Architectures and Trusted Network Connect

Active Open-Source Software Development

- xmhf.org. Hypervisor boot integrity measurement (including dynamic root of trust support on AMD and Intel platforms), cryptographic key management, TrustVisor API and micro-TPM.

- flickertcb.sf.net. Isolated execution for security-sensitive code on x86-class systems from AMD and Intel with support for dynamic root of trust, and is compatible with Linux and Windows.

Selected Publications

1. Bauer, L., Garriss, S., McCune, J.M., Reiter, M.K., Rouse, J., Rutenbar, P.: Device-enabled authorization in the grey system. In: Proceedings of the Information Security Conference, July 2005
2. Chen, C.-H.O., Chen, C.-W., Kuo, C., Lai, Y.-H., McCune, J.M., Studer, A., Perrig, A., Yang, B.-Y., Wu, T.-C.: GAnGS: gather, authenticate, and group securely. In: Proceedings of the International Conference on Mobile Computing and Networking (Mobicom), September 2008
3. Filyanov, A., McCune, J.M., Sadeghi, A.-R., Winandy, M.: Uni-directional trusted path: transaction confirmation on just one device. In: Proceedings of the IEEE/IFIP International Conference on Dependable Systems and Networks (DSN), June 2011
4. Franklin, J., Chaki, S., Datta, A., McCune, J.M., Vasudevan, A.: Para-metric verification of address space separation. In: Proceedings of the Conference on Principles of Security and Trust (POST), March 2012
5. Li, Y., McCune, J.M., Perrig, A.: VIPER: verifying the integrity of PERipherals' firmware. In: Proceedings of the ACM Conference on Computer and Communications Security (CCS), October 2011
6. Libonati, A., McCune, J.M., Reiter, M.K.: Usability testing a Malware-resistant input mechanism. In: Proceedings of the Network and Distributed System Security Symposium (NDSS), February 2011
7. Lin, Y.-H., Studer, A., Hsiao, H.-C., McCune, J.M., Wang, K.-H., Krohn, M., Lin, P.-L., Perrig, A., Sun, H.-M., Yang, B.-Y.: SPATE: small-group PKI-less authenticated trust establishment. In: Proceedings of the Conference on Mobile Systems, Applications and Services (MobiSys), June 2009 (Best Paper Award)
8. McCune, J.M., Berger, S., Cáceres, R., Jaeger, T., Sailer, R.: Shamon: a system for distributed mandatory access control. In: Proceedings of the Annual Computer Security Applications Conference (ACSAC), December 2006
9. McCune, J.M., Li, Y., Qu, N., Zhou, Z., Datta, A., Gligor, V., Perrig, A.: TrustVisor: effcient TCB reduction and attestation. In: Proceedings of the IEEE Symposium on Security and Privacy, May 2010
10. McCune, J.M., Parno, B., Perrig, A., Reiter, M.K., Seshadri, A.: Minimal TCB code execution (extended abstract). In: Proceedings of the IEEE Symposium on Security and Privacy, May 2007
11. McCune, J.M., Parno, B., Perrig, A., Reiter, M.K., Isozaki, H.: Flicker: an execution infrastructure for TCB minimization. In: Proceedings of the European Conference on Computer Systems (EuroSys), April 2008

12. McCune, J.M., Parno, B., Perrig, A., Reiter, M.K., Seshadri, A.: How low can you go? recommendations for hardware-supported minimal TCB code execution. In: Proceedings of the Architectural Support for Programming Languages and Operating Systems (ASPLOS), March 2008
13. McCune, J.M., Perrig, A., Reiter, M.K.: Seeing is believing: using camera phones for human-verifiable authentication. In: Proceedings of the IEEE Symposium on Security and Privacy, May 2005
14. McCune, J.M., Perrig, A., Reiter, M.K.: Safe passage for passwords and other sensitive data. In: Proceedings of the Network and Distributed System Security Symposium (NDSS), February 2009
15. McCune, J.M., Perrig, A., Seshadri, A., van Doorn, L.: Turtles all the way down: research challenges in user-based attestation. In: Proceedings of the USENIX Workshop on Hot Topics in Security (HotSec '07) (2007)
16. Parno, B., Lorch, J.R., Douceur, J.R., Mickens, J., McCune, J.M.: Memoir: practical state continuity for protected modules. In: Proceedings of the IEEE Symposium on Security and Privacy, May 2011
17. Parno, B., McCune, J.M., Perrig, A.: Bootstrapping trust in commodity computers. In: Proceedings of the IEEE Symposium on Security and Privacy, May 2010
18. Schwartz, E.J., Brumley, D., McCune, J.M.: A contractual anonymity system. In: Proceedings of the Network and Distributed System Security Symposium (NDSS), February 2010
19. Vasudevan, A., McCune, J.M., Newsome, J., Perrig, A., van Doorn, L.: CARMA: a hardware tamper-resistant isolated execution environment on commodity $\times 86$ plat-forms. In: Proceedings of the ACM Symposium on Information, Computer and Communications Security (ASIACCS), May 2012
20. Vasudevan, A., Owusu, E., Zhou, Z., Newsome, J., McCune, J.M.: Trustworthy execution on mobile devices: what security properties can my mobile platform give me? In: Proceedings of the International Conference on Trust and Trustworthy Computing (Trust), June 2012
21. Zhou, Z., Gligor, V.D., Newsome, J., McCune, J.M.: Building verifiable trusted path on commodity $\times 86$ computers. In: Proceedings of the IEEE Symposium on Security and Privacy, May 2012

Selected Technical Reports

1. Chaki, S., Vasudevan, A., Jia, L., McCune, J., Datta, A.: Design, development, and automated verification of an integrity-protected hypervisor. Technical Report CMU-CyLab-12-017, CyLab, Carnegie Mellon University, Pittsburgh, PA, July 2012
2. Vasudevan, A., McCune, J.M., Newsome, J.: Design and implementation of an eXtensible and modular hypervisor framework. Technical report CMU-CyLab-12-014, Cylab, Carnegie Mellon University, Pittsburgh, PA, June 2012

Patents

1. McCune, J.M., Gligor, V.D., Perrig, A., Datta, A., Parno, B.J., Qu, N., Vasudevan, A., Li, Y.: User-verifiable execution of security-sensitive code on untrusted
   platforms. Serial No. 12/720,008, pending since June 2010
2. McCune, J.M., Perrig, A., Datta, A., Gligor, V.D., Qu, N.: Methods and appa-
   ratuses for user-verifiable trusted path in the presence of Malware. International
   Patent PCT/US2010/040334, WIPO No. WO 2011/037665, pending since June
   2010

Professsional Experience

Co-Founder and President; Software Engineer (February 2010–Present)
NoFuss Security, Inc, Pittsburgh, PA USA

- Member of three-person engineering team: design, development, and support for commercialization of trustworthy computing technologies. Projects
  include ongoing design and development for the Flicker system (Linux and
  Windows driver development, AMD CPU microcode loading, Intel VT-d
  DMA-protection mechanisms, $\times 86$ virtual memory), and analysis of COTS
  hypervisor security.

Software Engineer (May 2010–April 2012)
VDG, Inc, Pittsburgh, PA USA

- Phase II STTR, Army Research Offce (ARO) Topic A08-T005: Trustworthy Execution of Security-Sensitive Code on Untrusted Systems. Design
  and implementation of trustworthy computing functionality for the TrustVisor hypervisor on top of XMHF. Includes boot integrity of XMHF itself
  via dynamic root of trust and TPM, Micro-TPM API for code running in
  TrustVisor's protected environment, cryptographic key management, remote
  attestation protocols, sealed storage APIs, ensuring state continuity for all
  sensitive state, and regression testing infrastructure.

Consultant and Developer (2010–2011)
Wave Systems, Cupertino, CA USA

- Boot integrity, support for virtualization.

Consultant—Trusted Computing (February 2008–April 2008)
VMware Corporation, Palo Alto, CA USA

- Studied the applicability of emerging trusted computing technologies to virtualization.

Summer Intern—Systems Security (May 2005–August 2005)
IBM Research, Hawthorne, NY USA

- Designed, implemented, and analyzed an extension to the sHype hypervisor security architecture for the Xen hypervisor. This extension enables bridging of mandatory access control (MAC) enforcement between two physically separate systems.

Summer Intern—SDET (May 2002–August 2002)
Microsoft Corporation, Redmond, Washington USA

- Developed two performance benchmarking applications for WinFS based on analysis of customer profiles. Designed, developed, and deployed an application to automate benchmark installation, execution, and result analysis for a cluster of performance-analysis machines. Shared the responsibility of educating several full-time employees who were hired during my time at Microsoft.

Talks

- TrustVisor: Effcient TCB Reduction and Attestation (IEEE S&P, Oakland, CA, May 2010).
- Safe Passage for Passwords and Other Sensitive Data (NDSS, February 2009).
- How low can you go? Recommendations for Hardware-Supported Minimal TCB Code Execution (ASPLOS, March 2008).
- Shamon: A System for Distributed Mandatory Access Control (ACSAC, Miami Beach, FL, December 2006).
- Bump in the Ether: A Framework for Securing Sensitive User Input (Usenix ATC, Boston, MA, June 2006).
- Seeing is Believing: Using Camera Phones for Human-Verifiable Authentication (IEEE S&P, Oakland, CA, May 2005).

Professional Service

- PC Member: 2013 International Conference on Architectural Support for Programming Languages and Operating Systems (ASPLOS).
- PC Member: 2013 Network and Distributed System Security Symposium (NDSS).
- PC Member: 2012 USENIX Workshop on Hot Topics in Security (HotSec).
- PC Member: 2012 USENIX Security Symposium.
- PC Member: TRUST 2012: International Conference on Trust and Trustworthy Computing.

- PC Member: 2012 IEEE Symposium on Security and Privacy (Oakland).
- PC Member: 2011 Workshop on Scalable Trusted Computing (STC).
- General Chair: TRUST 2011: International Conference on Trust and Trustworthy Computing.
- PC Member: 2011 IEEE Symposium on Security and Privacy (Oakland)
- PC Member: 2011 International Workshop on Security and Privacy in Spontaneous Interaction and Mobile Phone Use.
- PC Member: 2010 Workshop on Scalable Trusted Computing (STC).
- PC Member: 2010 IEEE Symposium on Security and Privacy (Oakland).
- PC Member: 2010 International Workshop on Security and Privacy in Spontaneous Interaction and Mobile Phone Use.
- PC Member: TRUST 2010: International Conference on Trust and Trustworthy Computing.
- PC Member: 2009 Workshop on Scalable Trusted Computing (STC).

Honors and Awards

- Best paper award for SPATE at MobiSys (2009).
- Recipient of the A. G. Jordan Award, for combining outstanding Ph.D. thesis work with exceptional service to the ECE community (2009).
- Best paper award for GAnGS at MobiCom (2008).
- University of Virginia: graduated with High Distinction in Computer Engineering (2003).
- Honorable Mention: ACM Programming Contest World Finals (2003).

Computer Skills

- Languages: C/C++, $\times 86$ Assembly, Perl, Python, Java, shell
- Systems software: Linux, Xen, Apple OS X, Windows
- Low-level programming and tools

James Newsome

Personal Details

Name: James Newsome

E-mail: http://www.jimnewsome.net
             jim@jimnewsome.net

Education

M.S. in Electrical and Computer Engineering (December 2003)
Carnegie Mellon University, Pittsburgh, PA

Ph.D. in Electrical and Computer Engineering (December 2008)
Carnegie Mellon University, Pittsburgh, PA
GPA: 3.94

BSE in Computer Engineering (May 2002)
University of Michigan, Ann Arbor, MI
GPA: 3.88

Employment

Sole Proprietor (August 2012)
jimnewsome.net, Pittsburgh, PA

- I am an independent developer. I am available for freelance and consulting in cyber-security and software development.

Systems Scientist (July 2010–August 2012)
Carnegie Mellon University, Pittsburgh, PA

- Performed research and development in virtualization and trusted computing. I was especially involved in the development of the TrustVisor hypervisor, and the TEE-SDK. These tools allow a piece of code to run in isolation from the rest of the computer system, allowing security-critical Pieces of Application Code to run securely even if the operating system is infested with malware. We released these as open source as part of the xmhf project (http://xmhf.org).

Research Engineer (November 2008–July 2010)
Bosch Research and Technology Center Pittsburgh, PA

- Performed internal consulting on cryptography and design of secure networked embedded systems.

Intern Researcher (May 2004–January 2005, May 2005–August 2005)
Intel Research Pittsburgh, Pittsburgh, PA

- Researched and developed Polygraph, a system to automatically generate signatures for polymorphic worms. Work resulted in an open source release, and two publications (IEEE S&P 2005 and RAID 2006)

Peer-Reviewed Publications

1. Brumley, D., Caballero, J., Liang, Z., Newsome, J., Song, D.: Towards automatic discovery of deviations in binary implementations with applications to error detection and fingerprint generation. In: Proceedings of the 2007 USENIX Security Conference (2007)
2. Brumley, D., Newsome, J., Song, D., Wang, H., Jha, S.: Towards automatic generation of vulnerability-based signatures. In: Proceedings of the 2006 IEEE Symposium on Security and Privacy, May 2006
3. Kim, T.H.-J., Bauer, L., Newsome, J., Perrig, A., Walker, J.: Challenges in access right assignment for secure home networks. In: Proceedings of the 5th USENIX Workshop on Hot Topics in Security (HotSec '10) (2010)
4. Kim, T.H.-J., Bauer, L., Newsome, J., Perrig, A., Walker, J.: Access right assignment mechanisms for secure home networks. J. Commun. Networks **13**(2), 175–186 (2011)
5. Newsome, J., Brumley, D., Song, D.: Vulnerability-specific execution filtering for exploit prevention on commodity software. In: Proceedings of the 13th Annual Network and Distributed System Security Symposium (NDSS '06), February 2006
6. Newsome, J., Brumley, D., Franklin, J., Song, D.: Replayer: automatic protocol replay by binary analysis. In: Proceedings of the 13th ACM Conference on Computer and Communications Security (CCS), October 2006
7. Newsome, J., Karp, B., Song, D.: Olygraph: automatically generating signatures for polymorphic worms. In: Proceedings of the IEEE Symposium on Security and Privacy, May 2005
8. Newsome, J., Karp, B., Song, D.: Paragraph: thwarting signature learning by training maliciously. In: Proceedings of the 9th International Symposium On Recent Advances In Intrusion Detection (RAID 2006), September 2006
9. Newsome, J., McCamant, S., Song, D.: Measuring channel capacity to distinguish undue influence. In: Proceedings of the 4th ACM SIGPLAN Workshop on Programming Languages and Analysis for Security (PLAS), June 2009
10. Newsome, J., Shi, R., Song, D., Perrig, A.:The sybil attack in sensor networks: analysis & defenses. In: Proceedings of the 3rd International Symposium on Information Processing in Sensor Networks (IPSN '04), April 2004
11. Newsome, J., Song, D.: GEM: graph EMbedding for routing and data-centric storage in wireless sensor networks. In: Proceedings of ACM SenSys, November 2003
12. Newsome, J., Song, D.: Dynamic taint analysis for automatic detection, analysis, and signature generation of exploits on commodity software. In: Proceedings of the 12th Annual Network and Distributed System Security Symposium (NDSS '05), February 2005
13. Vasudevan, A., McCune, J.M., Newsome, J., Perrig, A., van DoornCARMA, L.: A hardware tamper-resistant isolated execution environment on commodity

×86 platforms. In: Proceedings of the ACM Symposium on Information, Computer and Communications Security (ASIACCS) (2012)

14. Vasudevan, A., Owusu, E., Zhou, Z., Newsome, J., McCune, J.M.: Trustworthy execution on mobile devices: what security properties can my mobile platform give me? In: Proceedings of Trust and Trustworthy Computing (2012)

15. Tucek, J., Newsome, J., Lu, S., Huang, C., Xanthos, S., Brumley, D., Zhou, Y., Song, D.: Sweeper: a lightweight end-to-end system for defending against fast worms. In: Proceedings of the 2nd ACM SIGOPS/EuroSys European Conference on Computer Systems (2007)

16. Zhou, Z., Gligor, V., Newsome, J., McCune, J.: Building verifiable trusted path on commodity ×86 computers. In: Proceedings of the IEEE Symposium on Security and Privacy (2012)

Technical Reports

1. Brumley, D., Hartwig, C., Kang, M.G., Liang, Z., Newsome, J., Poosankam, P., Song, D., Yin, H.: BitScope: automatically dissecting malicious binaries. Technical report CMU-CS-07-133, Carnegie Mellon School of Computer Science, March 2007

2. Brumley, D., Liang, Z., Newsome, J., Song, D.: Towards practical automatic generation of multipath vulnerabity signatures. Technical report CMU-CS-07-150, Carnegie Mellon University School of Computer Science (2007)

3. Brumley, D., Newsome, J.: Alias analysis for assembly. Technical report CMU-CS-06-180, Carnegie Mellon University School of Computer Science, December 2006

4. Newsome, J., Brumley, D., Song, D.: Sting: an end-to-end self-healing system for defending against zero-dayWorm attacks on commodity software. Technical report CMU-CS-05-191 (2005)

5. Newsome, J., Song, D.: Influence: a quantitative approach for data integrity. Technical report CMU-CyLab-08-005, Carnegie Mellon Cylab, February 2008

Book Chapters

1. Brumley, D., Hartwig, C., Liang, Z., Newsome, J., Poosankam, P., Song, D., Yin, H.: Automatically identifying trigger-based behavior in Malware. In: Lee, W., Wang, C., Dagon, D. (eds.) Botnet Analysis and Defense, vol. 36 of Advances in Information Security Series, pp. 65–88. Springer (2008)

2. Brumley, D., Newsome, J., Song, D.: Sting: an end-to-end self-healing system for defending against internet worms. In: Malware Detection. Springer (2007)